Photocatalytic Hydrogen Evolution

Photocatalytic Hydrogen Evolution

Special Issue Editors

Misook Kang
Vignesh Kumaravel

MDPI • Basel • Beijing • Wuhan • Barcelona • Belgrade • Manchester • Tokyo • Cluj • Tianjin

Special Issue Editors

Misook Kang

Vignesh Kumaravel

Yeungnam University

Institute of Technology Sligo

Korea

Ireland

Editorial Office

MDPI

St. Alban-Anlage 66

4052 Basel, Switzerland

This is a reprint of articles from the Special Issue published online in the open access journal *Catalysts* (ISSN 2073-4344) (available at: https://www.mdpi.com/journal/catalysts/special_issues/hydrogen_evolution).

For citation purposes, cite each article independently as indicated on the article page online and as indicated below:

LastName, A.A.; LastName, B.B.; LastName, C.C. Article Title. *Journal Name* **Year**, *Article Number, Page Range*.

ISBN 978-3-03936-310-0 (Pbk)

ISBN 978-3-03936-311-7 (PDF)

Cover image courtesy of Vignesh Kumaravel.

Contents

About the Special Issue Editors

Misook Kang (Professor) obtained her Ph.D. in energy and hydrocarbon chemistry from Kyoto University, Japan in 1998. She was a full-time Research Professor at Sungkyunkwan and KyungHee University in Korea from 1999 to 2005, and was involved in photocatalysis research work. Since 2006, she has been a professor at the school of chemistry and biochemistry of Yeungnam University in Gyeongbuk. Her research interests are in the area of renewable energy, particularly focused on hydrogen production from photo- and thermal-catalysis using various nanomaterials. She has, to date, published more than 250 papers on energy and environment-related topics in peer-reviewed SCI(E) journals. In addition, she has received numerous academic awards, including the Gyeongbuk Science and Technology Award (the Women's S & T Award) in 2015. She served as a member of the Editorial Board of the *Journal of Industrial Engineering and Chemistry* (JIEC) from 2008 to 2014. Currently, she is the Acting Financial Director in the Korean Society of Industrial and Engineering Chemistry (KSIEC).

Vignesh Kumaravel (Senior Research Fellow) obtained his Ph.D. in Chemistry from Madurai Kamaraj University, India in 2013. He later worked as a Research Professor at Yeungnam University, Republic of Korea. He was then awarded a post-doctoral fellowship for an industrial project at Universiti Sains Malaysia. In October 2016, he joined Texas A & M University, Qatar, as an Assistant Research Scientist. Since March 2018 Vignesh has been working at IT Sligo as a Senior Research Fellow in the Renewable Engine project. He has published several scientific research articles in international peer-reviewed journals and presented his research findings at several international conferences. He has also delivered various invited international talks in the Republic of Korea, Spain, India, etc. He is acting as a co-investigator for three major research grants sponsored by Malaysian funding agencies. He is also acting as a potential reviewer for many Elsevier, ACS, RSC and Wiley journals. To his credit, he has reviewed more than 50 research articles.

Editorial

Photocatalytic Hydrogen Evolution

Vignesh Kumaravel [1],* and Misook Kang [2],*

[1]   Department of Environmental Science, Institute of Technology Sligo, Ash Lane, Co., Sligo F91 YW50, Ireland
[2]   Department of Chemistry, College of Natural Sciences, Yeungnam University, Gyeongsan, Gyeongbuk 38541, Korea
*   Correspondence: Kumaravel.Vignesh@itsligo.ie (V.K.); mskang@ynu.ac.kr (M.K.)

Received: 21 April 2020; Accepted: 26 April 2020; Published: 1 May 2020

Solar energy conversion is one of the sustainable technologies that tackles the global warming and energy crisis. Photocatalytic hydrogen (H_2) production is a clean technology to produce eco-friendly fuel with the help of semiconductor nanoparticles and abundant sunlight irradiation. Titanium oxide (TiO_2), graphitic-carbon nitride (g-C_3N_4) and cadmium sulfide (CdS) are the most widely explored photocatalysts in recent decades for water splitting.

As the guest editors, we have comprehensively investigated the role of sacrificial agents on the H_2 production efficiency of TiO_2, g-C_3N_4 and CdS photocatalysts [1]. The activity of the catalysts was evaluated without any noble metal co-catalysts. The effects of the most widely reported sacrificial agents were evaluated in this work. The activity of the catalysts was influenced by the number of hydroxyl groups, alpha hydrogen and carbon chain length of the sacrificial agent. We found that glucose and glycerol are the most suitable sacrificial agents to produce H_2 with minimum toxicity to the solution. The findings of this study would be highly favorable for the selection of a suitable sacrificial agent for photocatalytic H_2 production.

Hong et al. demonstrated the photoelectrochemical (PEC) efficiency of $MoSe_2$/Si nanostructures for H_2 production and carbon dioxide (CO_2) reduction [2]. PEC deposition coupled with the rapid thermal annealing method was applied to fabricate the electrodes on the Si substrate. PEC H_2 evolution and CO_2 conversion efficiencies of the $MoSe_2$/Si electrode were higher in visible light irradiation as compared to dark conditions.

Kim et al. synthesised monodispersed spherical TiO_2 particles with a disordered rutile surface for photocatalytic H_2 production [3]. The photocatalyst was synthesised through sol-gel and a chemical reduction technique using Li/ethylenediamine (Li/EDA) solution. The samples were calcined at various temperatures to tune the anatase to the rutile phase ratio. The disordered rutile surface and mixed crystalline phase of TiO_2 significantly increased the H_2 production under solar light irradiation.

Idrees et al. reported the photocatalytic activity of Nb_2O_5/g-C_3N_4 heterostructures for molecular H_2 production under simulated solar light irradiation [4]. A hydrothermal technique was utilised to develop the three dimensional Nb_2O_5/g-C_3N_4 heterostructure with a high surface area. H_2 production efficiency of Nb_2O_5/g-C_3N_4 (10 wt. %) was higher than that of pure Nb_2O_5 and g-C_3N_4. The photogenerated electron hole pairs were successfully separated through a direct Z-scheme mechanism at the heterojunction.

Kim and Woodward described the band gap modulation of tantalum (V) perovskite by anion control [5]. Perovskites such as $BaTaO_2N$, $SrTaO_2N$, $CaTaO_2N$, $KTaO_3$, $NaTaO_3$ and TaO_2F were studied in this work. Pt-loaded $CaTaO_2N$ was utilised as a visible-light-driven photocatalyst for H_2 production using CH_3OH as the sacrificial agent.

Son et al. reported the impact of sulfur defects on the H_2 production efficiency of a CuS@$CuGaS_2$ heterojunction under visible light irradiation [6]. The activity of the CuS@$CuGaS_2$ heterojunction was higher as compared to pure CuS. This was ascribed to the introduction of structural defects to promote the photo-generated electron hole separation.

The recent accomplishments in the synthesis and application of various photocatalysts for H_2 production are briefly reviewed by Zhang et al. [7]

Tremendous efforts should be taken in the future to commercialise this photocatalytic technology at the industry level. The studies should also be performed with cheap materials, industrial wastewater and seawater for H_2 production.

Finally, we would like to convey our sincere thanks to all the authors for their significant contributions in this special issue.

Author Contributions: Conceptualization, V.K. and M.K.; Review and editing, V.K and M.K. All authors have read and agreed to the published version of the manuscript.

Conflicts of Interest: The authors declare no conflict of interest

References

1. Kumaravel, V.; Imam, M.D.; Badreldin, A.; Chava, R.K.; Do, J.Y.; Kang, M.; Abdel-Wahab, A. Photocatalytic hydrogen production: Role of sacrificial reagents on the activity of oxide, carbon, and sulfide catalysts. *Catalysts* **2019**, *9*, 276. [CrossRef]

2. Hong, S.; Rhee, C.K.; Sohn, Y. Photoelectrochemical Hydrogen Evolution and CO2 Reduction over MoS2/Si and MoSe2/Si Nanostructures by Combined Photoelectrochemical Deposition and Rapid-Thermal Annealing Process. *Catalysts* **2019**, *9*, 494. [CrossRef]

3. Kim, N.Y.; Lee, H.K.; Moon, J.T.; Joo, J.B. Synthesis of Spherical TiO2 Particles with Disordered Rutile Surface for Photocatalytic Hydrogen Production. *Catalysts* **2019**, *9*, 491. [CrossRef]

4. Idrees, F.; Dillert, R.; Bahnemann, D.; Butt, F.K.; Tahir, M. In-Situ Synthesis of Nb2O5/g-C3N4 Heterostructures as Highly Efficient Photocatalysts for Molecular H2 Evolution under Solar Illumination. *Catalysts* **2019**, *9*, 169. [CrossRef]

5. Kim, Y.-I.; Woodward, P.M. Band gap modulation of Tantalum (V) perovskite semiconductors by anion control. *Catalysts* **2019**, *9*, 161. [CrossRef]

6. Son, N.; Heo, J.N.; Youn, Y.-S.; Kim, Y.; Do, J.Y.; Kang, M. Enhancement of Hydrogen Productions by Accelerating Electron-Transfers of Sulfur Defects in the CuS@ CuGaS2 Heterojunction Photocatalysts. *Catalysts* **2019**, *9*, 41. [CrossRef]

7. Zhang, Y.; Heo, Y.-J.; Lee, J.-W.; Lee, J.-H.; Bajgai, J.; Lee, K.-J.; Park, S.-J. Photocatalytic hydrogen evolution via water splitting: A short review. *Catalysts* **2018**, *8*, 655. [CrossRef]

 *catalysts* 

Article

Photocatalytic Hydrogen Production: Role of Sacrificial Reagents on the Activity of Oxide, Carbon, and Sulfide Catalysts

Vignesh Kumaravel [1,2,*], Muhammad Danyal Imam [3], Ahmed Badreldin [3], Rama Krishna Chava [4], Jeong Yeon Do [4], Misook Kang [4,*] and Ahmed Abdel-Wahab [3,*]

[1] Department of Environmental Science, School of Science, Institute of Technology Sligo, Ash Lane, F91 YW50 Sligo, Ireland

[2] Centre for Precision Engineering, Materials and Manufacturing Research (PEM), Institute of Technology Sligo, Ash Lane, F91 YW50 Sligo, Ireland

[3] Chemical Engineering Program, Texas A&M University at Qatar, Doha 23874, Qatar; muhammad.imam@qatar.tamu.edu (M.D.I.); ahmed.badreldin@qatar.tamu.edu (A.B.)

[4] Department of Chemistry, College of Natural Sciences, Yeungnam University, Gyeongsan, Gyeongbuk 38541, Korea; drcrkphysics@hotmail.com (R.K.C.); daengi77@ynu.ac.kr (J.Y.D.)

[*] Correspondence: Kumaravel.Vignesh@itsligo.ie (V.K.); mskang@ynu.ac.kr (M.K.); ahmed.abdel-wahab@qatar.tamu.edu (A.A.-W.)

Received: 15 February 2019; Accepted: 11 March 2019; Published: 18 March 2019

Abstract: Photocatalytic water splitting is a sustainable technology for the production of clean fuel in terms of hydrogen (H_2). In the present study, hydrogen (H_2) production efficiency of three promising photocatalysts (titania (TiO_2-P25), graphitic carbon nitride (g-C_3N_4), and cadmium sulfide (CdS)) was evaluated in detail using various sacrificial agents. The effect of most commonly used sacrificial agents in the recent years, such as methanol, ethanol, isopropanol, ethylene glycol, glycerol, lactic acid, glucose, sodium sulfide, sodium sulfite, sodium sulfide/sodium sulfite mixture, and triethanolamine, were evaluated on TiO_2-P25, g-C_3N_4, and CdS. H_2 production experiments were carried out under simulated solar light irradiation in an immersion type photo-reactor. All the experiments were performed without any noble metal co-catalyst. Moreover, photolysis experiments were executed to study the H_2 generation in the absence of a catalyst. The results were discussed specifically in terms of chemical reactions, pH of the reaction medium, hydroxyl groups, alpha hydrogen, and carbon chain length of sacrificial agents. The results revealed that glucose and glycerol are the most suitable sacrificial agents for an oxide photocatalyst. Triethanolamine is the ideal sacrificial agent for carbon and sulfide photocatalyst. A remarkable amount of H_2 was produced from the photolysis of sodium sulfide and sodium sulfide/sodium sulfite mixture without any photocatalyst. The findings of this study would be highly beneficial for the selection of sacrificial agents for a particular photocatalyst.

Keywords: photocatalysis; TiO_2; g-C_3N_4; CdS; energy

---

1. Introduction

Photocatalytic hydrogen (H_2) production via water splitting is a sustainable and renewable energy production technology with negligible impact on the environment [1] (Figure 1). H_2 is one of the most promising and clean energy sources for the future, with water as the only combustion product. After the invention of photo-electrochemical water splitting in 1972 [2] by Fujishima and Honda, nearly 9000 research articles have been published, outlining the use of various photocatalysts. In particular, most of the research works have been carried out using powder photocatalysts (except photo-electrochemical studies). The reported materials in the recent years are categorized as oxide [3–149], carbon [3,81,150–237], and sulfide [3,14,17,35,58,59,113,114,119,128,133,154,164,169,177,181,

195,203,208,210,215,220,227,230,235,238–345] photocatalysts. Titanium oxide–P25 (TiO_2-P25), graphitic carbon nitride (g-C_3N_4), and cadmium sulfide (CdS) are the most extensively studied photocatalysts for water splitting. Many review articles have also been published [1,116,163,167,225,237,238,245,346–419] discussing the various features of the photocatalytic water splitting, such as fundamental concepts, theoretical principles, nature (morphology, surface characteristics, and optical properties) of the photocatalyst, role of co-catalyst/sacrificial reagents, mechanism, kinetics, etc. Nevertheless, there is still not many comprehensive studies to identify an appropriate sacrificial reagent with respect to the nature of a photocatalyst.

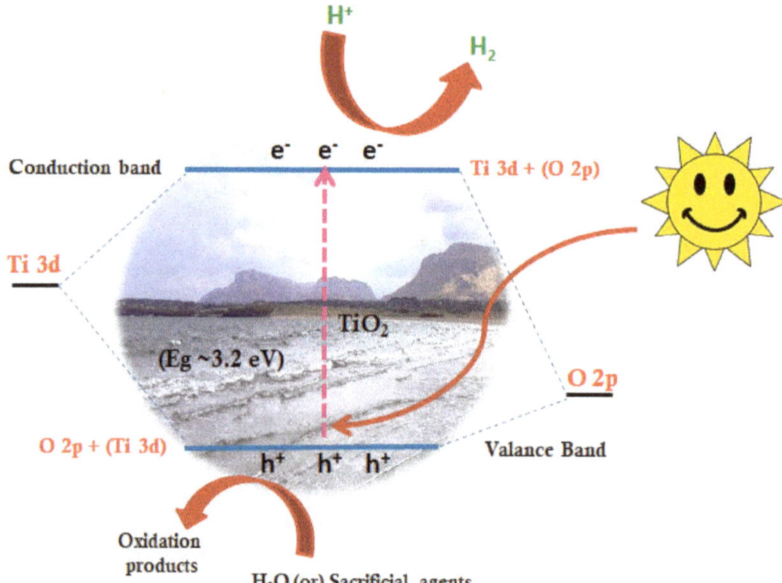

Figure 1. Schematic representation of the water-splitting process on a photocatalyst surface under light irradiation [1]. Reproduced with permission from Ref. [1]. Copyright 2019, Elsevier.

Sacrificial agents or electron donors/hole scavengers play a prominent role in photocatalytic H_2 production because the water splitting is energetically an uphill reaction (ΔH_0 = 286 kJ mol^{-1}). It is realized that methanol, triethanolamine, and sodium sulfide/sodium sulfite are the most commonly used sacrificial reagents for oxide, carbon, and sulfide photocatalysts, respectively. In most of the cases, fresh water (e.g., deionized water or double distilled water) has been used to evaluate the H_2 production efficiency in a micro photo-reactor (volume in the range of 30 to 70 mL) with a strong light irradiation source (nearly $\leq$ 300 W). However, the vitality and utilization of this technology have not been comprehensively studied in a real environment. Moreover, the commercialization of this technology is still restrained by its poor efficiency and the use of expensive noble metals (like Pt, Au, Pd, Rh) as co-catalysts. Most of the published results do not have much consistency in terms of efficiency. For example, different efficiency values have been reported for pure TiO_2 using methanol as a scavenger (Table 1). This discrepancy is ascribed to the following reasons: photo-reactor design, inert gas (Ar or N_2) purging flow rate, light irradiation source, gas sampling method, gas chromatography (GC) analysis conditions, calculations, etc.

Table 1. Photocatalytic H_2 production efficiency of TiO_2 using methanol sacrificial agent.

Catalyst Amount (g/L)	Concentration of Methanol (%)	Light Source	H_2 Production Efficiency (µmol/g/h)	Reference
1	10	300 W of Xe (without UV cutoff filter)	42.00	[420]
0.6	16.66	300 W of Xe (with UV cutoff filter)	18.47	[217]
0.5	20	300 W of Xe (with UV cutoff filter)	~20.00	[194]
1.29	25.8	300 W of Xe (with UV cutoff filter)	~2.00	[421]

The photochemical reactions of sacrificial agents (methanol, ethanol, isopropanol, ethylene glycol, glycerol, glucose, lactic acid, triethanolamine, sodium sulfide, sodium sulfite, and sodium sulfide/sodium sulfite mixture) and their degradation products during H_2 production are summarized as follows:

Methanol [422] (MeOH):

$$H_2O_{(l)} + h^+ \rightarrow {}^\bullet OH + H^+ \tag{1}$$

$$CH_3OH_{(l)} + {}^\bullet OH \rightarrow {}^\bullet CH_2OH + H_2O_{(l)} \tag{2}$$

$${}^\bullet CH_2OH \rightarrow HCHO_{(l)} + H^+ + e^- \tag{3}$$

$$2H^+ + 2e^- \rightarrow H_{2\,(g)} \tag{4}$$

$$HCHO_{(l)} + H_2O_{(l)} \rightarrow HCOOH_{(l)} + H_{2\,(g)} \tag{5}$$

$$HCOOH_{(l)} \rightarrow CO_{2\,(g)} + H_{2\,(g)} \tag{6}$$

Overall reaction:

$$CH_3OH_{(l)} + H_2O_{(l)} \rightarrow CO_{2\,(g)} + 3H_{2\,(g)} \tag{7}$$

Ethanol [423] (EtOH):

$$CH_3CH_2OH + TiO_2 \rightarrow {}_{(s)}CH_3CH_2O - Ti^{4+} + {}_{(s)}OH \tag{8}$$

$$TiO_2 + UV\ light \rightarrow 2e^-_{(a)} + 2h^+ \tag{9}$$

$${}_{(s)}CH_3CH_2O - Ti^{4+} + 2h^+ \rightarrow {}_{(s)}CH_3CHO + Ti^{4+} \tag{10}$$

$${}_{(s)}2OH + e^-_{(a)} \rightarrow H_2 + {}_{(s)}2O^{2+} \tag{11}$$

Here, (s) represents the photocatalyst surface and (a) denotes the photo-excited electrons by UV light.

Isopropanol [424] (IPA):

$$\left[{}^{S^2}_{Cd^{2+}} > (CdS) \right]_2 + H_2O \rightarrow {}^{S^2}_{Cd^{2+}} > Cd(II)SH + {}^{S^2}_{Cd^{2+}} > S(-II)Cd(II)OH \tag{12}$$

$${}^{S^{2-}}_{Cd^{2+}} > CdSH_2^+ \rightarrow {}^{S^{2-}}_{Cd^{2+}} > CdSH + H^+ \tag{13}$$

$${}^{S^{2-}}_{Cd^{2+}} > CdSH \rightarrow {}^{S^{2-}}_{Cd^{2+}} > CdS - + H^+ \tag{14}$$

$${}^{S^{2-}}_{Cd^{2+}} > CdOH_2^+ \rightarrow {}^{S^{2-}}_{Cd^{2+}} > CdOH + H^+ \tag{15}$$

$${}^{S^{2-}}_{Cd^{2+}} > CdOH \rightarrow {}^{S^{2-}}_{Cd^{2+}} > CdO - + H^+ \tag{16}$$

$$\underset{Cd^{2+}}{S^2} > Cd(+II)S(0)^+ + C_3H_7OH \rightarrow \underset{Cd^{2+}}{S^2} > Cd(+II)S(-I)H^+ + C_3H_6^{\bullet}OH \tag{17}$$

$$\underset{Cd^{2+}}{S^{2-}} > Cd(+II)S(-I)H^+ + C_3H_7OH \rightarrow \underset{Cd^{2+}}{S^{2-}} > Cd(+II)S(-II)H_2^+ + C_3H_6^{\bullet}OH \tag{18}$$

$$2H^{\bullet} \rightarrow H_2 \tag{19}$$

$$2\,C_3H_6^{\bullet}OH \rightarrow 2\,C_3H_5O + H_2 \tag{20}$$

$$\underset{Cd^{2+}}{S^2} > Cd(+II)S(-II)H_2^+ \rightarrow \underset{Cd^{2+}}{S^2} > Cd(+II)S(-II)H + H^+ \tag{21}$$

$$\underset{Cd^{2+}}{S^{2-}} > CdOH + e_{CB}^- \rightarrow \underset{Cd^{2+}}{S^{2-}} > CdO^- + H^+ \tag{22}$$

Ethylene Glycol [76,425] (EG):

$$OHCH_2 - CH_2OH + H_2O \xrightarrow{TiO_2,\ h\nu} OHCH_2 - CHO \tag{23}$$

$$OHCH_2 - CHO \xrightarrow{\bullet OH} OHCH_2 - COOH \tag{24}$$

$$OHCH_2 - COOH \rightarrow CH_3COOH \tag{25}$$

$$OHCH_2 - COOH \rightarrow HOOC - COOH \tag{26}$$

$$HOOC - COOH \rightarrow HCOOH \tag{27}$$

$$HCOOH\ (or)\ CH_3COOH\ (or)\ HOOC - COOH \rightarrow CO_2 + H_2 + CH_4 + C_2H_4 + C_2H_6 + H_2O \tag{28}$$

Glycerol [130] (GLY):

$$C_3H_8O_3 + 3H_2O + 14\,h_{(VB)}^+ \rightarrow \text{intermediates}\ (C_2H_4O_2,\ C_2H_2O_3,\ C_2H_4O_3,\ C_3H_6O_3,\ etc)$$
$$\rightarrow 3CO_2 + 14\,H^+ \tag{29}$$

$$14H^+ + 14e_{CB}^- \rightarrow 7H_2\ (g) \tag{30}$$

Glucose [9] (GLU):

$$C_6H_{12}O_6 + H_2O\ (anaerobic) \rightarrow C_5H_{10}O_5 + HCOOH + H_{2(g)} \tag{31}$$

$$C_5H_{10}O_5 + H_2O \rightarrow C_4H_8O_4 + HCOOH + H_{2\ (g)} \tag{32}$$

$$C_4H_8O_4 + H_2O + HCOOH + H_{2\ (g)}\,(aerobic) \rightarrow HCOOH + H_{2(g)} + CO_{2\ (g)} \tag{33}$$

$$C_6H_{12}O_6 \xrightarrow{TiO_2,\ h\nu, H_2O,\ O_2} C_6H_{12}O_7 \tag{34}$$

$$C_6H_{12}O_7 \xrightarrow{TiO_2,\ h\nu, H_2O,\ O_2} C_6H_{10}O_8 \tag{35}$$

$$C_6H_{10}O_8 \xrightarrow{TiO_2,\ h\nu, H_2O,\ O_2} HCOOH + H_{2(g)} + CO_{2(g)} \tag{36}$$

Lactic Acid [426] (LA):

$$CH_3 - CH(OH) - COOH + H_2O \xrightarrow{TiO_2,\ h\nu} CO_2 + H_2 + CH_3 - CO - COOH \tag{37}$$

Triethanolamine [427] (TEOA):

$$C_6H_{15}NO_3 \rightarrow C_6H_{15}NO_3^+ + e^- \tag{38}$$

$$C_6H_{15}NO_3^+ \rightarrow C_6H_{14}NO_3^{\bullet} + H^+ \tag{39}$$

$$C_6H_{14}NO_3^{\bullet} \rightarrow C_6H_{14}NO_3^+ + e^- \tag{40}$$

$$C_6H_{14}NO_3^+ + H_2O \rightarrow C_4H_{11}NO_3 + CH_3CHO + H^+ \tag{41}$$

Sodium sulfide (Na₂S) [428]:

$$Na_2S + H_2O \rightarrow 2Na^+ + S^{2-} \tag{42}$$

$$S^{2-} + H_2O \rightarrow HS^- + OH^- \tag{43}$$

$$HS^- + h\nu \rightarrow HS^{-*} \tag{44}$$

$$HS^{-*} + HS^- \rightarrow [(HS)_2]^{-*} \rightarrow H_2 + S_2^{2-} \tag{45}$$

Sodium sulfite (Na₂SO₃) [429]:

$$\text{Irradiation}: SO_3^{2-} \xrightarrow{h\nu} SO_3^{2-*} \tag{46}$$

$$\text{Oxidation}: SO_3^{2-*} + 2OH^- \rightarrow SO_4^{2-} + H_2O + 2e^- \tag{47}$$

$$\text{Reduction}: 2H_2O + 2e^- \rightarrow H_2 + 2OH^- \tag{48}$$

$$\text{Oxidation}: 2SO_3^{2-} \rightarrow S_2O_6^{2-} + 2e^- \tag{49}$$

$$\text{Reduction}: 2H_2O + 2e^- \rightarrow H_2 + 2OH^- \tag{50}$$

Sodium sulfide and sodium sulfite mixture (Na₂S and Na₂SO₃) [430]:

Two different reaction pathways are involved when sodium sulfide and sodium sulfite mixture is used as a sacrificial agent.

$$HS^-_{(aq)} \rightarrow HS^-_{(ads)} \tag{51}$$

$$HS^-(ads) \xrightarrow{h\nu} [HS^-(ads)]^* \tag{52}$$

Path A:

$$[HS^-(ads)]^* \rightarrow H^* + S^-(ads) \tag{53}$$

$$S^-(ads) + [HS^-(ads)]^* \rightarrow [HS_2^{2-}]\varphi \tag{54}$$

$$[HS_2^{2-}]\varphi \rightarrow H^* + S_2^{2-}(ads) \tag{55}$$

$$2H^* \rightarrow H_2 \tag{56}$$

Path B:

$$[HS^-(ads)]^* + S^0(ads) \rightarrow [HS_2^-]\varphi \tag{57}$$

$$[HS_2^-]\varphi + OH^- + SO_3^{2-} + S^0(ads) \rightarrow [HS_2^{2-}]\varphi + OH^\bullet + S_2O_3^{2-} \tag{58}$$

$$OH^\bullet + SO_3^{2-} \xrightarrow{h\nu} SO_4^{2-} + H^* \tag{59}$$

$$2H^* \rightarrow H_2 \tag{60}$$

$$[HS^-(ads)]^* + H_2O \rightarrow S^0(ads) + H_2 + OH^- \tag{61}$$

$$[HS_2^-]\varphi + OH^- \rightarrow H_2O + S_2^{2-} \tag{62}$$

where (ads) denotes adsorption and φ represents species, which can undergo intramolecular charge transfer.

The previous articles reported H_2 production efficiencies with various combinations of photocatalysts and sacrificial reagents. This study provides detailed information on the selection of sacrificial reagents and photocatalysts for H_2 production. The efficiencies of TiO₂-P25, g-C₃N₄, and CdS were evaluated using methanol (MeOH), ethanol (EtOH), isopropanol (IPA), ethylene glycol

(EG), glycerol (GLY), lactic acid (LA), glucose (GLU), sodium sulfide (Na_2S), sodium sulfite (Na_2SO_3), sodium sulfide/sodium sulfite mixture (Na_2S/Na_2SO_3), and triethanolamine (TEOA) as sacrificial reagents (organic and inorganic). The efficiency of a photocatalyst was described in terms of pH of medium and nature of the sacrificial agent (carbon chain length, alpha hydrogen, hydroxyl groups, binding interactions, etc). Besides, control experiments were executed to investigate the H_2 production with only sacrificial reagents under solar light irradiation in the absence of photocatalyst.

2. Results and Discussion

2.1. TiO$_2$ P25

Figure 2 shows the H_2 production efficiency of TiO_2 P25 using various sacrificial agents. H_2 production efficiencies of TiO_2/EG, TiO_2/GLY, TiO_2/Na_2S/Na_2SO_3, TiO_2/GLU, TiO_2/Na_2S were found to be 190.2 µmol, 130.8 µmol, 126 µmol, 120 µmol, and 120 µmol, respectively. H_2 production efficiency of TiO_2/MeOH system reduced to 81.6 µmol for the same period. The use of TEOA, EtOH, IPA, and Na_2SO_3 as sacrificial reagents resulted in poor H_2 production, yielding 61.8 µmol, 49.8 µmol, 46.2 µmol, and 40.8 µmol, respectively. TiO_2/LA mixture displayed the lowest yield of H_2 production (only 27.6 µmol). TiO_2/EG mixture showed the maximum H_2 production (190.2 µmol) efficiency as compared to all other combinations. This is ascribed to the faster charge transfer reaction in the TiO_2/EG system compared to the photo-generated electron-hole recombination process [431,432]. The length of the carbon chain, the number of hydroxyl groups, and dehydrogenation/decarbonylation characteristics of sacrificial agents are the primary features in controlling the H_2 production efficiency. Moreover, the following properties of sacrificial agents could also strongly influence the efficiency: polarity and electron donating ability, adsorption capability on the photocatalyst surface, the formation of by-products, and the selectivity for reaction with photo-generated holes (e.g., decarboxylation process) [10,94,431–436]. Carbon monoxide (CO) is one of the main intermediates for the alcohols with a short carbon chain. Hence, the adsorption of CO on the active sites of TiO_2 via chemisorption restricts further adsorption of alcohol on the photocatalyst surface [437]. The removal of CO as CO_2 is the rate-determining step in H_2 production. It depends on the adsorption efficiency and the number of alpha hydrogens of the sacrificial agent [437]. During the water-splitting process, the hydroxyl radical ($^{\bullet}$OH) abstracts alpha hydrogen from the alcohol to create $^{\bullet}RCH_2$-OH radical, which gets further oxidized into an aldehyde, carboxylic acid, and CO_2 [437]. Bahruji et al. [437] suggested that alkyl groups connected to the alcohol (e.g., C_xH_yOH) could yield the respective alkanes (e.g., C_{x-1}) during the water-splitting process. The alkane production rate was decreased with the increase of OH groups in alcohol [438]. In the case of polyols, the hydrogen atoms from the alpha carbon could be easily extracted and evolved in the form of H_2 [438]. The alpha carbon atoms could be oxidized into CO_2. The C atoms without OH groups (other than alpha C atoms) would be evolved in the form of alkanes [438]. Time-resolved transient absorption spectroscopy results revealed that carbohydrates and polyols (C2–C6) could rapidly react with ~50–60% holes (h^+) within 6 ns as compared to other alcohols [439,440]. The OH groups could act as an anchor for the chemisorption of alcohols on the photocatalyst surface [438]. The coordination efficiency of alcohols with the Ti sites relies on the number of OH groups and the carbon chain length. This type of linkage could be beneficial for the utilization of holes to improve the H_2 production and suppress the charge carrier recombination [438]. The first principle calculations showed that the formation of gap levels in TiO_2 via the adsorption polyols could accelerate the hole trapping process [441]. Though EG showed maximum efficiency for TiO_2-P25, glycerol and glucose are the most appropriate sacrificial agents for any kind of oxide photocatalyst. This owes to their (glucose and glycol) most abundance, less toxicity, low cost, and they can readily undergo dehydrogenation as compared to other alcohols [40,63,435].

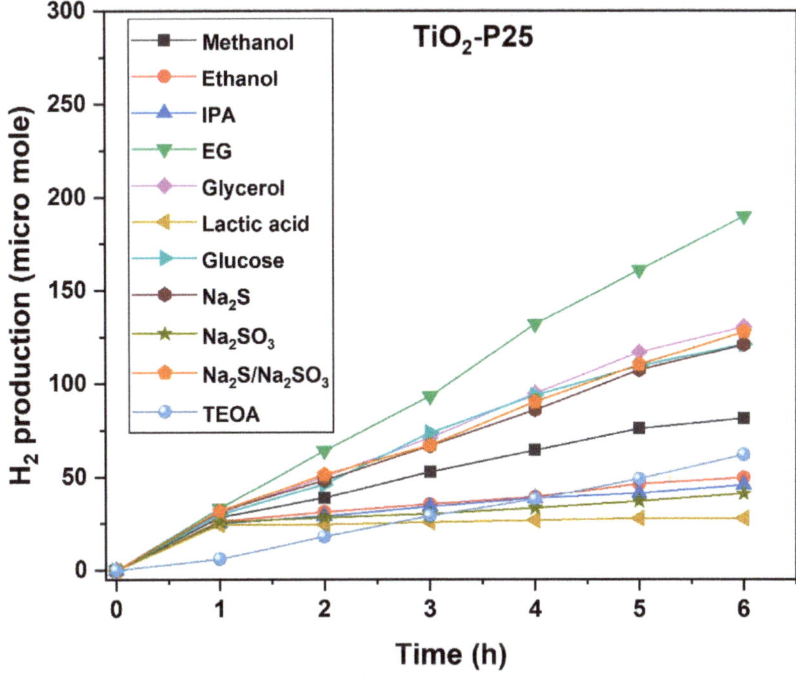

Figure 2. Photocatalytic H$_2$ production efficiency of TiO$_2$-p25 using various sacrificial agents.

2.2. *g*-C$_3$N$_4$

H$_2$ production efficiency of *g*-C$_3$N$_4$ with various sacrificial agents is shown in Figure 3. In this case, only the use of TEOA, Na$_2$S, Na$_2$SO$_3$, and Na$_2$S/Na$_2$SO$_3$ resulted in H$_2$ production. H$_2$ production efficiency of *g*-C$_3$N$_4$/Na$_2$S (139.8 μmol) system was higher than that of *g*-C$_3$N$_4$/Na$_2$S/Na$_2$SO$_3$ (127.2 μmol) and *g*-C$_3$N$_4$/Na$_2$SO$_3$ (5.4 μmol). *g*-C$_3$N$_4$/TEOA mixture showed the best efficiency (247.2 μmol) when compared to all other sacrificial agents. This can be ascribed to the fact that photo-corrosion and degradation of π conjugated structure [304] of amine rich *g*-C$_3$N$_4$ is secured by the effective binding of TEOA on the catalyst surface [112]. TEOA excellently consumes the photo-generated holes, improves the dispersion of photocatalyst, and acts as a binding ligand to improve the interaction of *g*-C$_3$N$_4$ with water molecules [204,442]. The results shown in Figure 3 also suggest that alcohols and glucose are not strongly adsorbed on the *g*-C$_3$N$_4$ surface for water-splitting reaction. This is attributed to the absence of hydrophilicity and surface characteristics (e.g., active sites, poor electrical conductivity, water oxidation ability) of *g*-C$_3$N$_4$ to facilitate a strong interfacial electron/hole transfer process on the catalyst surface. The poor crystallinity and basal planar structure of *g*-C$_3$N$_4$ endorse the electron-hole recombination [443]. Moreover, high activation energy and overpotential are required for H$_2$ production on the *g*-C$_3$N$_4$ surface [182,211]. This could be rectified by the loading of noble metals or co-catalysts over *g*-C$_3$N$_4$ or fabricating Z-scheme photocatalysts. In most of the studies, it was reported that *g*-C$_3$N$_4$ acts as an outstanding template and there was no H$_2$ production on *g*-C$_3$N$_4$ without any noble metal co-catalyst [444,445]. The results also demonstrated that the light absorption capability, chemical stability, and suitable band edge positions of narrow band-gap *g*-C$_3$N$_4$ are not the only decisive factors to enhance the H$_2$ production efficiency.

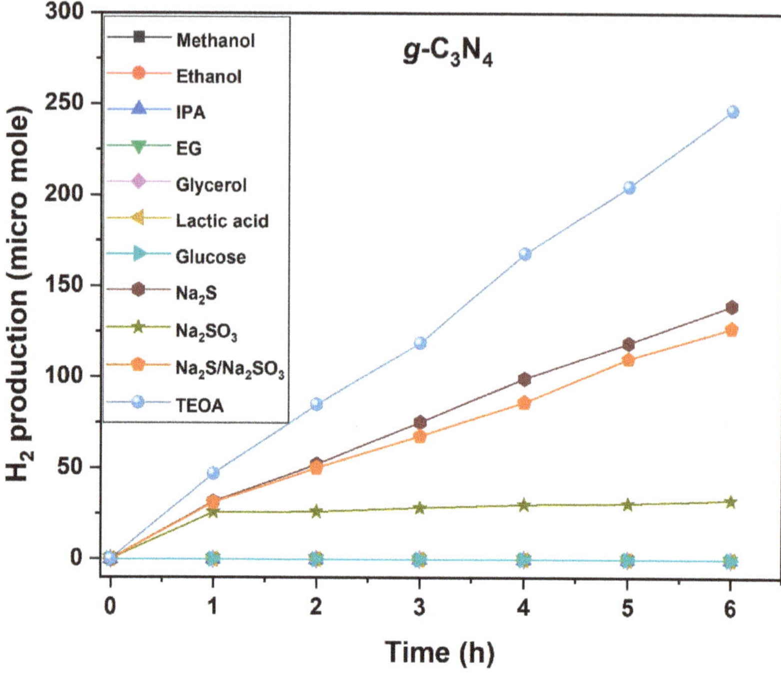

Figure 3. Photocatalytic H$_2$ production efficiency of g-C$_3$N$_4$ using various sacrificial agents.

2.3. CdS

Photocatalytic H$_2$ production efficiency of CdS using various sacrificial agents is shown in Figure 4. The use of TEOA, Na$_2$S, Na$_2$SO$_3$, Na$_2$S/Na$_2$SO$_3$, and LA as sacrificial reagents resulted in H$_2$ formation. CdS/TEOA system showed the maximum efficiency of 283.2 µmol of H$_2$ as compared to all other sacrificial agents. The efficiency of CdS/Na$_2$S, CdS/Na$_2$SO$_3$, CdS/LA systems was found to be 181.2 µmol, 154.8 µmol, and 84 µmol, respectively. The mixture of CdS/Na$_2$S/Na$_2$SO$_3$ showed the lowest H$_2$ production of 54 µmol after 6 h. Bare CdS is not stable under prolonged light irradiation because the sulfide ions on its surface are rapidly oxidized into sulfur through the reaction with photo-generated holes (photo-corrosion – CdS + 2h$^+$ → Cd^{2+} + S) [308,446,447]. The sulfide oxidation of CdS can occur before the oxidation of water by holes [308,447]. Hence, the H$_2$ production efficiency of CdS highly relies on the effective binding of sacrificial agents on its surface. The results showed that amine and sulfide/sulfite might be strongly bound to the CdS surface and it could effectively consume the holes as compared to alcohol and sugars. It is obviously noted that H$_2$ is produced in high alkaline (amine, sulfide, and sulfite) and acidic (LA) pH mixtures when compared to neutral pH (alcohols and sugar). LA is converted into pyruvic acid and CO$_2$ during the water-splitting reaction; this may slightly influence the pH and polarity of the reaction mixture. The sulfide ions from Na$_2$S stabilizes CdS surface to terminate the surface defects originated from photo-corrosion. The electron-hole recombination process is strongly restrained by the sulfide ions at alkaline pH. When CdS is suspended in a water medium, thiol (Cd-SH) and hydroxyl (Cd-OH) groups are developed on its surface, which are highly pH dependent [310]. In the case of Na$_2$S, the pH of the medium is alkaline, sulfide (S$_2^-$) and hydrogen sulfide (HS$^-$) are formed when Na$_2$S is dissolved in water [310]. During light irradiation, S$_2^-$ and HS$^-$ are quickly oxidized into sulfate (SO$_4^{2-}$) and polysulfide (S$_4^{2-}$, S$_5^{2-}$) ions, respectively [310]. The oxidation of sulfide by the photo-generated holes is much preferential as compared to the photo-corrosion of CdS [112]. The precipitation of yellow colored polysulfide ions diminishes the photocatalytic efficiency via acting as an optical filter and competing with the H_2 generation reaction. This could be restricted by the addition of Na_2SO_3 to generate more HS^- and $S_2O_3^{2-}$ ions to enhance the photocatalytic activity [292]. However, the results shown in Figure 4 suggest that H_2 production efficiency of Na_2S/CdS or CdS/Na_2SO_3 are higher than that of $CdS/Na_2S/Na_2SO_3$. The reasons could be predicted by the photolysis experiments of sacrificial agents. The pH of TEOA/water mixture would be around 12, which could enhance the H_2 production efficiency via strong interfacial bonding on the CdS surface and its reaction with photo-generated holes [204].

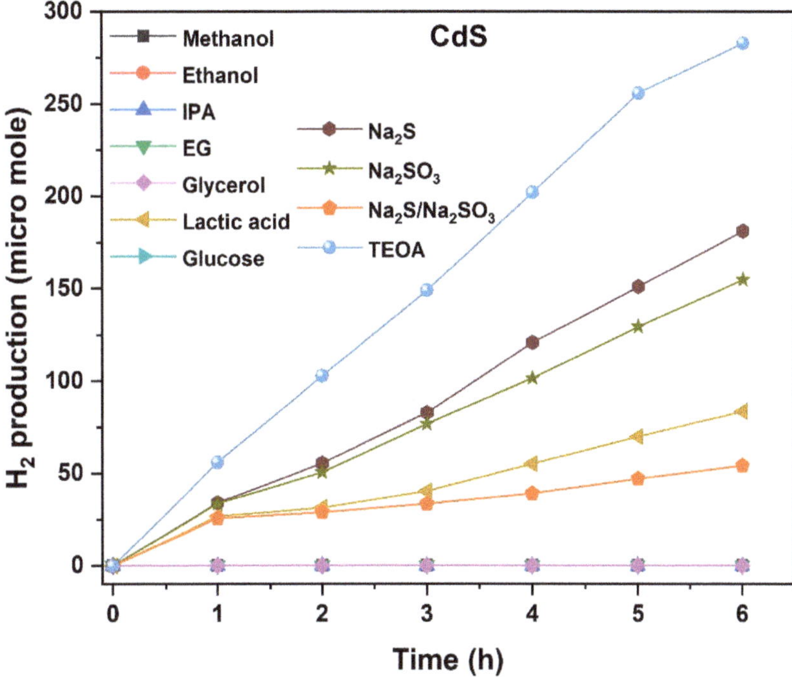

Figure 4. Photocatalytic H_2 production efficiency of CdS using various sacrificial agents.

2.4. Photolysis

Photolysis experiments were carried out for all sacrificial agents in water for 6 h of light irradiation without the additions of photocatalysts. Control experiments were also carried out in the absence of sacrificial agents to evaluate the efficiency of the photocatalyst. There was no H_2 production in the absence of any sacrificial agents for TiO_2-p25, g-C_3N_4, and CdS. The results of photolysis experiments with sacrificial reagents under solar light in the absence of photocatalysts are shown in Figure 5. Interestingly, a remarkable amount of H_2 was evolved from Na_2S/water (159 μmol), Na_2SO_3/water (51 μmol), and Na_2S/Na_2SO_3/water (134.4 μmol) systems without photocatalyst. It was observed that the H_2 production efficiency was increased with respect to the concentration of sulfide or sulfite. When compared to results obtained in the presence of photocatalysts, it could be observed that the photocatalysts, such as TiO_2-p25 and g-C_3N_4, surprisingly reduced the actual H_2 production efficiency of sulfide system. There was not a significant increment in the efficiency of CdS/Na_2S as compared to the photolysis of Na_2S. However, the efficiency of CdS/Na_2SO_3 was higher than that of Na_2SO_3 photolysis. It is also noted that a high concentration of sulfide/sulfite mixture (in the range of 0.2 M to 1 M) was used in most of the studies for H_2 production [246,264,273,300,448–451]. In such cases, the photolysis of sulfide or sulfite solutions were not evaluated. Hence, the H_2 production should have been mainly originated via photolysis of sulfide/sulfite mixture rather than the photocatalytic effect. The photochemical reactions involved in H_2 generation from sulfide and sulfite solutions are described in detail from Equations (42)–(62) [428–430].

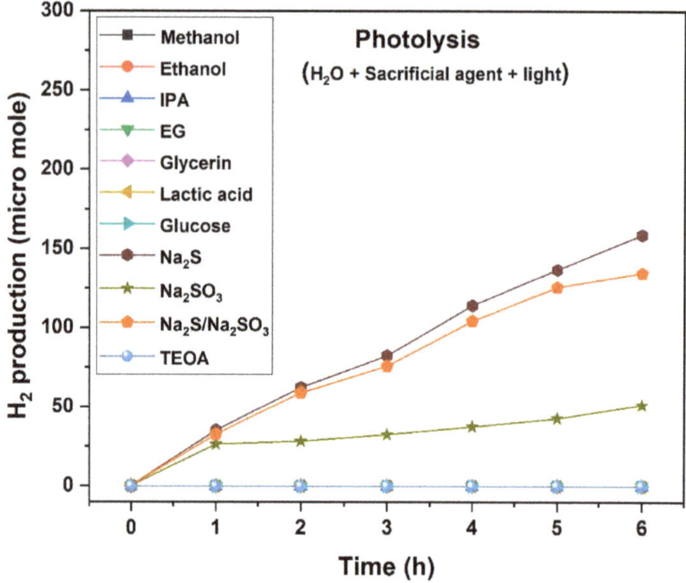

Figure 5. H_2 production efficiency of sacrificial agents without photocatalyst (photolysis).

Li et al. [430] investigated the photochemical generation of H_2 from sulfide and sulfite mixture solutions. They found that the pH of sulfide/sulfite mixture (~13.14) was slightly decreased (~12.94) after the completion of photolysis experiments. The addition of a small amount of sulfite into the sulfide solution could not amplify the H_2 production. Nevertheless, the elemental sulfur or polysulfide originated from HS^- is efficiently consumed by SO_3^{2-} to improve the photonic efficiency. It is strongly recommended to study the effect of photolysis when the sulfide/sulfite mixture is used as the sacrificial agent to evaluate the H_2 production efficiency of the photocatalyst. The elemental sulfur could be filtered out by passing hydrogen sulfide (H_2S) gas in the aqueous solution under nitrogen atmosphere [428]. The resulting filtrate could be reused for photolytic H_2 production [428].

Wang et al. [112] suggested that Na_2S/Na_2SO_3, MeOH, and TEOA were the most appropriate sacrificial agents for sulfide, oxide, and carbon-based photocatalysts. However, the experiments were not performed to study the effect of photolysis, especially the sulfide or sulfite solutions. A high concentration of sacrificial agents was used to evaluate the photocatalytic activity. The photocatalytic experiments were not described in detail. Moreover, the effect of most earth abundant glucose and glycerol were not investigated on the H_2 production efficiency. Sulfur dioxide (SO_2) emission from the flue gas can be absorbed as Na_2SO_3 solution using dilute sodium hydroxide. The photolysis of such sodium sulfite solution is an eco-friendly way to produce H_2 gas [429]. Only a few studies were focused on using wastewater for H_2 generation. Souza and Silva [310] studied the feasibility of using tannery sludge wastewater for photocatalytic H_2 generation using CdS. The photolysis of sulfide-rich wastewater or industrial effluent is the foremost choice to produce green energy in a sustainable way via photocatalysis.

2.5. TOC Analysis

TOC analysis was carried out for solutions after the photocatalytic experiments, to assess the degradation of organic sacrificial agents in the solutions at the end of 6 h experiment. Table 2 summarizes the TOC results for the solutions of TiO_2-p25. The results suggest the effective utilization of the organic sacrificial agents by TiO_2-P25 for H_2 production. TOC was reduced almost half for most of the sacrificial agents after 6 h except for glucose. This is ascribed to the formation of more organic intermediates when glucose is utilized as the sacrificial agent.

Table 2. TOC of solutions after the completion of the photocatalytic reaction.

Sample	TOC (mg/L)	
	Blank (Before Light Irradiation)	**TiO_2-P25**
Water	8.659	-
Methanol	30,450	15,530
Ethanol	43,680	17,070
Isopropanol	55,010	17,770
Glycerol	54,220	18,700
Ethylene glycol	59,080	17,930
Glucose	7699	11,500
Lactic acid	47,310	20,440

3. Experimental

All the chemicals used were of analytical grade and used as received without further purification. TiO_2 P25 was purchased from Sigma Aldrich (Darmstadt, Germany), g-C_3N_4 was synthesized by the calcination of urea at 550 °C for 2 h [452]. CdS was synthesized via the hydrothermal method [453]. Photocatalytic experiments were carried out using an immersion type reactor (Lelesil innovative systems, Thane, Maharashtra, India) as shown in Figure 6. All the reactions were carried out without any noble metal co-catalyst. The reactor is a tightly closed setup with a total volume of 1000 mL. Reactions were carried out using 500 mL of double distilled water with 0.5 g/L of photocatalyst and desired amount of sacrificial reagent (based on the literature). The empty headspace was kept constant at 500 mL for all reactions. The sacrificial agent concentration was fixed at 10% (alcohol, amine, acid) and 0.1 M (glucose, sodium sulfide, sodium sulfite, sodium sulfide/sodium sulfite mixture). The mixture was stirred under nitrogen purging for 1 h after which, the purging was stopped, and the reactor was closed immediately. The mixture was irradiated using a 300 W Xenon arc lamp without any UV cutoff filter (simulated solar light source). H_2 sampling was carried out for every 1 h using a 250 μL sample lock gas tight syringe. At the end of the photoreaction time, when organic sacrificial reagents were used, the mixture was filtered using a 0.45 μm micro-filter, and the filtrate was analyzed using a total organic carbon (TOC; Shimadzu, Japan) analyzer to measure the loss of reagents by mineralization. H_2 was analyzed using Agilent gas chromatography (USA) with thermal conductivity detector (TCD), manual injection, carrier gas N_2, molecular sieve 5 A° column with 2-m length, front inlet temperature 140 °C, and detector temperature 150 °C.

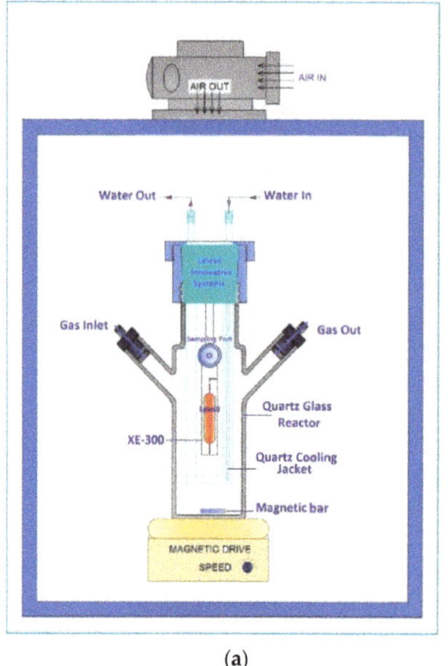
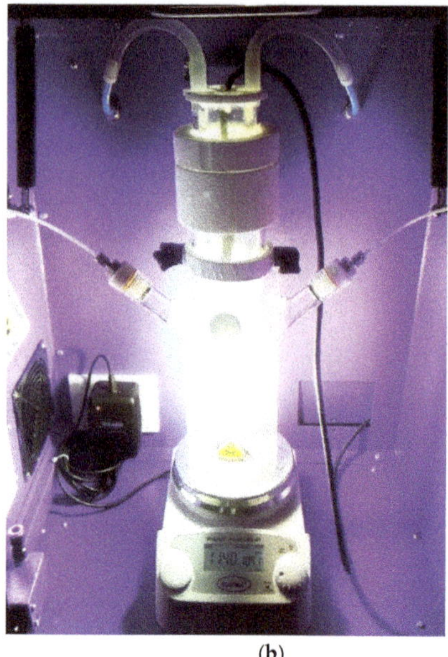

(a)   (b)

Figure 6. (**a**) Schematic and (**b**) photograph of the reactor used for photocatalytic H_2 production.

4. Summary and Outlook

Different types of photocatalysts have been successfully investigated for H_2 production using various organic and inorganic sacrificial agents. The surface of an oxide photocatalyst would be more suitable for polyols and sugars for adsorption as compared to amines and sulfides. Amines are the most appropriate of sacrificial agents for carbon and sulfide photocatalysts. CdS/TEOA (283.2 µmol), g-C_3N_4/TEOA (247.2 µmol), and TiO_2/EG (190.2 µmol) are the three best systems with maximum H_2 production in this study. H_2 could also be generated via the direct photolysis of sodium sulfide solution in the absence of any catalyst. TiO_2-p25 and g-C_3N_4 suppress the self-H_2 generation efficiency of Na_2S photolysis. More technological developments are required for the practical application of water-splitting in a scalable and economically feasible way. Stable, affordable, and active co-catalysts should be developed in the future to replace the expensive noble metals to achieve a significant amount of H_2 production. In most of the studies, precious fresh water with a high concentration of sacrificial agents was used in a small reactor to generate H_2. Hence, future studies should be focused on a pilot scale using industrial wastewater and seawater rather than using fresh water.

Author Contributions: Conceptualization V.K. and A.A-W.; Supervision, V.K., A.A-W. and M.K.; Methodology, design, investigation, data creation and analysis, V.K.; Methodology, data analysis, and original draft preparation, M.D.I and A.B.; Methodology and data analysis, J.Y.D. and R.K.C.

Funding: The authors are grateful to Texas A&M University at Qatar and Qatar Foundation for financial support.

Conflicts of Interest: The authors declare no conflict of interest.

References

1. Kumaravel, V.; Mathew, S.; Bartlett, J.; Pillai, S.C. Photocatalytic hydrogen production using metal doped TiO_2: A review of recent advances. *Appl. Catal. B Environ.* **2019**, *244*, 1021–1064. [CrossRef]
2. Fujishima, A.; Honda, K.J. Electrochemical photolysis of water at a semiconductor electrode. *Nature* **1972**, *238*, 37–38. [CrossRef] [PubMed]

3. Abdullah, H.; Kuo, D.-H.; Chen, X. High efficient noble metal free Zn (O, S) nanoparticles for hydrogen evolution. *Int. J. Hydrogen Energy* **2017**, *42*, 5638–5648. [CrossRef]

4. Agegnehu, A.K.; Pan, C.-J.; Tsai, M.-C.; Rick, J.; Su, W.-N.; Lee, J.-F.; Hwang, B.-J. Visible light responsive noble metal-free nanocomposite of V-doped TiO_2 nanorod with highly reduced graphene oxide for enhanced solar H_2 production. *Int. J. Hydrogen Energy* **2016**, *41*, 6752–6762. [CrossRef]

5. Alharbi, A.; Alarifi, I.M.; Khan, W.S.; Asmatulu, R. Synthesis and analysis of electrospun $SrTiO_3$ nanofibers with NiO nanoparticles shells as photocatalysts for water splitting. In Proceedings of the 14th Brazilian Polymer Conference, São Paulo, Brazil, 22–26 October 2017; 22-26.

6. Al-Mayman, S.I.; Al-Johani, M.S.; Mohamed, M.M.; Al-Zeghayer, Y.S.; Ramay, S.M.; Al-Awadi, A.S.; Soliman, M.A. TiO_2 ZnO photocatalysts synthesized by sol–gel auto-ignition technique for hydrogen production. *Int. J. Hydrogen Energy* **2017**, *42*, 5016–5025. [CrossRef]

7. Bai, Y.; Chen, T.; Wang, P.; Wang, L.; Ye, L. Bismuth-rich $Bi_4O_5X_2$ (X = Br, and I) nanosheets with dominant {101} facets exposure for photocatalytic H_2 evolution. *Chem. Eng. J.* **2016**, *304*, 454–460. [CrossRef]

8. Barreca, D.; Carraro, G.; Gasparotto, A.; Maccato, C.; Warwick, M.E.; Toniato, E.; Gombac, V.; Sada, C.; Turner, S.; Van Tendeloo, G. Iron–Titanium Oxide Nanocomposites Functionalized with Gold Particles: From Design to Solar Hydrogen Production. *Adv. Mater. Interfaces* **2016**, *3*, 1600348. [CrossRef]

9. Bellardita, M.; García-López, E.I.; Marcì, G.; Palmisano, L. Photocatalytic formation of H_2 and value-added chemicals in aqueous glucose (Pt)-TiO_2 suspension. *Int. J. Hydrogen Energy* **2016**, *41*, 5934–5947. [CrossRef]

10. Beltram, A.; Romero-Ocana, I.; Jaen, J.J.D.; Montini, T.; Fornasiero, P. Photocatalytic valorization of ethanol and glycerol over TiO_2 polymorphs for sustainable hydrogen production. *Appl. Catal. A Gen.* **2016**, *518*, 167–175. [CrossRef]

11. Betzler, S.B.; Podjaski, F.; Beetz, M.; Handloser, K.; Wisnet, A.; Handloser, M.; Hartschuh, A.; Lotsch, B.V.; Scheu, C. Titanium Doping and Its Effect on the Morphology of Three-Dimensional Hierarchical Nb_3O_7 (OH) Nanostructures for Enhanced Light-Induced Water Splitting. *Chem. Mater* **2016**, *28*, 7666–7672. [CrossRef]

12. Cargnello, M.; Montini, T.; Smolin, S.Y.; Priebe, J.B.; Jaén, J.J.D.; Doan-Nguyen, V.V.; McKay, I.S.; Schwalbe, J.A.; Pohl, M.-M.; Gordon, T.R. Engineering titania nanostructure to tune and improve its photocatalytic activity. *Proc. Natl. Acad. Sci.* **2016**, *113*, 3966–3971. [CrossRef]

13. Cha, G.; Altomare, M.; Truong Nguyen, N.; Taccardi, N.; Lee, K.; Schmuki, P. Double-Side Co-Catalytic Activation of Anodic TiO_2 Nanotube Membranes with Sputter-Coated Pt for Photocatalytic H_2 Generation from Water/Methanol Mixtures. *Chem. Asian J.* **2017**, *12*, 314–323. [CrossRef]

14. Yuan, Q.; Liu, D.; Zhang, N.; Ye, W.; Ju, H.; Shi, L.; Long, R.; Zhu, J.; Xiong, Y. Noble-Metal-Free Janus-like Structures by Cation Exchange for Z-Scheme Photocatalytic Water Splitting under Broadband Light Irradiation. *Angew. Chem.* **2017**, *129*, 4270–4274. [CrossRef]

15. Chen, L.; Gu, Q.; Hou, L.; Zhang, C.; Lu, Y.; Wang, X.; Long, J. Molecular p–n heterojunction-enhanced visible-light hydrogen evolution over a N-doped TiO_2 photocatalyst. *Catal. Sci. Technol.* **2017**, *7*, 2039–2049. [CrossRef]

16. Chen, W.; Liu, H.; Li, X.; Liu, S.; Gao, L.; Mao, L.; Fan, Z.; Shangguan, W.; Fang, W.; Liu, Y. Polymerizable complex synthesis of $SrTiO_3$:(Cr/Ta) photocatalysts to improve photocatalytic water splitting activity under visible light. *Appl. Catal. B Environ.* **2016**, *192*, 145–151. [CrossRef]

17. Xiang, Q.; Cheng, F.; Lang, D. Hierarchical Layered WS_2/Graphene-Modified CdS Nanorods for Efficient Photocatalytic Hydrogen Evolution. *ChemSusChem* **2016**, *9*, 996–1002. [CrossRef] [PubMed]

18. Chen, Z.; Wu, Y.; Wang, Q.; Wang, Z.; He, L.; Lei, Y.; Wang, Z. Oxygen-rich carbon-nitrogen quantum dots as cocatalysts for enhanced photocatalytic H_2 production activity of TiO_2 nanofibers. *Prog. Nat. Sci. Mater. Int.* **2017**, *37*, 333–337. [CrossRef]

19. Clarizia, L.; Vitiello, G.; Luciani, G.; Di Somma, I.; Andreozzi, R.; Marotta, R. In situ photodeposited nanoCu on TiO_2 as a catalyst for hydrogen production under UV/visible radiation. *Appl. Catal. A Gen.* **2016**, *518*, 142–149. [CrossRef]

20. Dhanalaxmi, K.; Yadav, R.; Kundu, S.K.; Reddy, B.M.; Amoli, V.; Sinha, A.K.; Mondal, J. $MnFe_2O_4$ Nanocrystals Wrapped in a Porous Organic Polymer: A Designed Architecture for Water-Splitting Photocatalysis. *Chem. Eur. J.* **2016**, *22*, 15639–15644. [CrossRef] [PubMed]

21. Ding, J.; Ming, J.; Lu, D.; Wu, W.; Liu, M.; Zhao, X.; Li, C.; Yang, M.; Fang, P. Study of the enhanced visible-light-sensitive photocatalytic activity of Cr_2O_3-loaded titanate nanosheets for Cr (vi) degradation and H_2 generation. *Catal. Sci. Technol.* **2017**, *7*, 2283–2297. [CrossRef]

22. Elbanna, O.; Kim, S.; Fujitsuka, M.; Majima, T. TiO_2 mesocrystals composited with gold nanorods for highly efficient visible-NIR-photocatalytic hydrogen production. *Nano Energy* **2017**, *35*, 1–8. [CrossRef]
23. Escobedo, S.; Serrano, B.; Calzada, A.; Moreira, J.; de Lasa, H. Hydrogen production using a platinum modified TiO_2 photocatalyst and an organic scavenger. Kinetic modeling. *Fuel* **2016**, *181*, 438–449. [CrossRef]
24. Feng, N.; Liu, F.; Huang, M.; Zheng, A.; Wang, Q.; Chen, T.; Cao, G.; Xu, J.; Fan, J.; Deng, F. Unravelling the Efficient Photocatalytic Activity of Boron-induced Ti3+ Species in the Surface Layer of TiO_2. *Sci. Rep.* **2016**, *6*, 34765. [CrossRef] [PubMed]
25. Fiorenza, R.; Bellardita, M.; D'Urso, L.; Compagnini, G.; Palmisano, L.; Scirè, S. Au/TiO_2-CeO_2 Catalysts for Photocatalytic Water Splitting and VOCs Oxidation Reactions. *Catalysts* **2016**, *6*, 121. [CrossRef]
26. Fontelles-Carceller, O.; Muñoz-Batista, M.J.; Conesa, J.C.; Fernández-García, M.; Kubacka, A. UV and visible hydrogen photo-production using Pt promoted Nb-doped TiO_2 photo-catalysts: Interpreting quantum efficiency. *Appl. Catal. B Environ.* **2017**, *216*, 133–145. [CrossRef]
27. Fornari, A.M.D.; de Araujo, M.B.; Duarte, C.B.; Machado, G.; Teixeira, S.R.; Weibel, D.E. Photocatalytic reforming of aqueous formaldehyde with hydrogen generation over TiO_2 nanotubes loaded with Pt or Au nanoparticles. *Int. J. Hydrogen Energy* **2016**, *41*, 11599–11607. [CrossRef]
28. Ge, M.; Cai, J.; Iocozzia, J.; Cao, C.; Huang, J.; Zhang, X.; Shen, J.; Wang, S.; Zhang, S.; Zhang, K.-Q. A review of TiO_2 nanostructured catalysts for sustainable H_2 generation. *Int. J. Hydrogen Energy* **2017**, *42*, 8418–8449. [CrossRef]
29. Guerrero-Araque, D.; Acevedo-Peña, P.; Ramírez-Ortega, D.; Calderon, H.A.; Gómez, R. Charge transfer processes involved in photocatalytic hydrogen production over CuO/ZrO_2–TiO_2 materials. *Int. J. Hydrogen Energy* **2017**, *42*, 9744–9753. [CrossRef]
30. Gullapelli, S.; Scurrell, M.S.; Valluri, D.K. Photocatalytic H_2 production from glycerol–water mixtures over Ni/γ-Al_2O_3 and TiO_2 composite systems. *Int. J. Hydrogen Energy* **2017**, *42*, 15031–15043. [CrossRef]
31. Gupta, B.; Melvin, A.A.; Matthews, T.; Dash, S.; Tyagi, A. TiO_2 modification by gold (Au) for photocatalytic hydrogen (H_2) production. *Renew. Sustain. Energy Rev.* **2016**, *58*, 1366–1375. [CrossRef]
32. Gusain, R.; Singhal, N.; Singh, R.; Kumar, U.; Khatri, O.P. Ionic-Liquid-Functionalized Copper Oxide Nanorods for Photocatalytic Splitting of Water. *ChemPlusChem* **2016**, *81*, 489–495. [CrossRef]
33. Hashimoto, T.; Ohta, H.; Nasu, H.; Ishihara, A. Preparation and photocatalytic activity of porous Bi_2O_3 polymorphisms. *Int. J. Hydrogen Energy* **2016**, *41*, 7388–7392. [CrossRef]
34. He, G.-L.; Zhong, Y.-H.; Chen, M.-J.; Li, X.; Fang, Y.-P.; Xu, Y.-H. One-pot hydrothermal synthesis of $SrTiO_3$-reduced graphene oxide composites with enhanced photocatalytic activity for hydrogen production. *J. Mol. Catal. A Chem.* **2016**, *423*, 70–76. [CrossRef]
35. Su, J.; Zhang, T.; Wang, L.; Shi, J.; Chen, Y. Surface treatment effect on the photocatalytic hydrogen generation of CdS/ZnS core-shell microstructures. *Chin. J. Catal.* **2017**, *38*, 489–497. [CrossRef]
36. Hinojosa-Reyes, M.; Hernández-Gordillo, A.; Zanella, R.; Rodríguez-González, V. Renewable hydrogen harvest process by hydrazine as scavenging electron donor using gold TiO_2 photocatalysts. *Catal. Today* **2016**, *266*, 2–8. [CrossRef]
37. Hou, H.-J.; Zhang, X.-H.; Huang, D.-K.; Ding, X.; Wang, S.-Y.; Yang, X.-L.; Li, S.-Q.; Xiang, Y.-G.; Chen, H. Conjugated microporous poly (benzothiadiazole)/TiO_2 heterojunction for visible-light-driven H_2 production and pollutant removal. *Appl. Catal. B Environ.* **2017**, *203*, 563–571. [CrossRef]
38. Hu, J.; Wang, L.; Zhang, P.; Liang, C.; Shao, G. Construction of solid-state Z-scheme carbon-modified TiO_2/WO_3 nanofibers with enhanced photocatalytic hydrogen production. *J. Power Sources* **2016**, *328*, 28–36. [CrossRef]
39. Hu, X.; Xiao, L.; Jian, X.; Zhou, W. Synthesis of mesoporous silica-embedded TiO_2 loaded with Ag nanoparticles for photocatalytic hydrogen evolution from water splitting. *J. Wuhan Univ. Technol. Mater. Sci. Ed.* **2017**, *32*, 67–75. [CrossRef]
40. Iervolino, G.; Vaiano, V.; Sannino, D.; Rizzo, L.; Palma, V. Enhanced photocatalytic hydrogen production from glucose aqueous matrices on Ru-doped $LaFeO_3$. *Appl. Catal. B Environ.* **2017**, *207*, 182–194. [CrossRef]
41. Jeong, S.; Chung, K.-H.; Lee, H.; Park, H.; Jeon, K.-J.; Park, Y.-K.; Jung, S.-C. Enhancement of Hydrogen Evolution from Water Photocatalysis Using Liquid Phase Plasma on Metal Oxide-Loaded Photocatalysts. *ACS Sustain. Chem. Eng.* **2017**, *5*, 3659–3666. [CrossRef]

42. Jiang, C.; Lee, K.Y.; Parlett, C.M.; Bayazit, M.K.; Lau, C.C.; Ruan, Q.; Moniz, S.J.; Lee, A.F.; Tang, J. Size-controlled TiO$_2$ nanoparticles on porous hosts for enhanced photocatalytic hydrogen production. *Appl. Catal. A Gen.* **2016**, *521*, 133–139. [CrossRef]

43. Jiang, Q.; Li, L.; Bi, J.; Liang, S.; Liu, M. Design and Synthesis of TiO$_2$ Hollow Spheres with Spatially Separated Dual Cocatalysts for Efficient Photocatalytic Hydrogen Production. *Nanomaterials* **2017**, *7*, 24. [CrossRef] [PubMed]

44. Jung, M.; Hart, J.N.; Boensch, D.; Scott, J.; Ng, Y.H.; Amal, R. Hydrogen evolution via glycerol photoreforming over Cu–Pt nanoalloys on TiO$_2$. *Appl. Catal. A Gen.* **2016**, *518*, 221–230. [CrossRef]

45. Jung, M.; Hart, J.N.; Scott, J.; Ng, Y.H.; Jiang, Y.; Amal, R. Exploring Cu oxidation state on TiO$_2$ and its transformation during photocatalytic hydrogen evolution. *Appl. Catal. A Gen.* **2016**, *521*, 190–201. [CrossRef]

46. Kang, H.W.; Park, S.B. Effects of Mo sources on Mo doped SrTiO$_3$ powder prepared by spray pyrolysis for H$_2$ evolution under visible light irradiation. *Mater. Sci. Eng. B* **2016**, *211*, 67–74. [CrossRef]

47. Kang, H.W.; Park, S.B. Improved performance of tri-doped photocatalyst SrTiO$_3$: Rh/Ta/F for H$_2$ evolution under visible light irradiation. *Int. J. Hydrogen Energy* **2016**, *41*, 13970–13978. [CrossRef]

48. Khan, M.; Sinatra, L.; Oufi, M.; Bakr, O.M.; Idriss, H. Evidence of plasmonic induced photocatalytic hydrogen production on Pd/TiO$_2$ upon deposition on thin films of gold. *Catal. Lett.* **2017**, *147*, 811–820. [CrossRef]

49. Khore, S.K.; Tellabati, N.V.; Apte, S.K.; Naik, S.D.; Ojha, P.; Kale, B.B.; Sonawane, R.S. Green sol–gel route for selective growth of 1D rutile N–TiO$_2$: A highly active photocatalyst for H$_2$ generation and environmental remediation under natural sunlight. *RSC Adv.* **2017**, *7*, 33029–33042. [CrossRef]

50. Kim, Y.K.; Seo, H.-J.; Kim, S.; Hwang, S.-H.; Park, H.; Lim, S.K. Effect of ZnO Electrodeposited on Carbon Film and Decorated with Metal Nanoparticles for Solar Hydrogen Production. *J. Mater. Sci. Technol.* **2016**, *32*, 1059–1065. [CrossRef]

51. Kumar, D.P.; Reddy, N.L.; Karthik, M.; Neppolian, B.; Madhavan, J.; Shankar, M. Solar light sensitized p-Ag$_2$O/n-TiO$_2$ nanotubes heterojunction photocatalysts for enhanced hydrogen production in aqueous-glycerol solution. *Sol. Energy Mater. Sol. Cells* **2016**, *154*, 78–87. [CrossRef]

52. Li, C.; Wang, H.; Lu, D.; Wu, W.; Ding, J.; Zhao, X.; Xiong, R.; Yang, M.; Wu, P.; Chen, F. Visible-light-driven water splitting from dyeing wastewater using Pt surface-dispersed TiO$_2$-based nanosheets. *J. Alloys Compd.* **2017**, *699*, 183–192. [CrossRef]

53. Li, F.; Wangyang, P.; Zada, A.; Humayun, M.; Wang, B.; Qu, Y. Synthesis of hierarchical Mn$_2$O$_3$ microspheres for photocatalytic hydrogen production. *Mater. Res. Bull.* **2016**, *84*, 99–104. [CrossRef]

54. Li, M.; Chen, Y.; Li, W.; Li, X.; Tian, H.; Wei, X.; Ren, Z.; Han, G. Ultrathin Anatase TiO$_2$ Nanosheets for High-Performance Photocatalytic Hydrogen Production. *Small* **2017**, *13*, 1604115. [CrossRef]

55. Li, R.; Zhao, Y.; Li, C. Spatial distribution of active sites on a ferroelectric PbTiO$_3$ photocatalyst for photocatalytic hydrogen production. *Faraday Discuss.* **2017**, *198*, 463–472. [CrossRef]

56. Li, X.; Gao, Y.; Liu, J.; Yu, X.; Li, Z. Facile synthesis of Ti3+ doped Ag/AgI TiO$_2$ nanoparticles with efficient visible-light photocatalytic activity. *Int. J. Hydrogen Energy* **2017**, *42*, 13031–13038. [CrossRef]

57. Liang, J.; Chai, Y.; Li, L. Facile fabrication of rod-shaped Zn$_2$GeO$_4$ nanocrystals as photocatalyst for hydrogen production. *Cryst. Res. Technol.* **2017**, *52*, 1700022. [CrossRef]

58. Kuang, P.Y.; Zheng, P.X.; Liu, Z.Q.; Lei, J.L.; Wu, H.; Li, N.; Ma, T.Y. Embedding Au quantum dots in rimous cadmium sulfide nanospheres for enhanced photocatalytic hydrogen evolution. *Small* **2016**, *12*, 6735–6744. [CrossRef]

59. Peng, B.; Liu, X.; Li, R. Preparation of a Single-Walled Carbon Nanotube/Cd0.8Zn0.2S Nanocomposite and Its Enhanced Photocatalytic Hydrogen Production Activity. *Eur. J. Inorg. Chem.* **2016**, *2016*, 3204–3212. [CrossRef]

60. Liu, Y.; Xiong, J.; Yang, Y.; Luo, S.; Zhang, S.; Li, Y.; Liang, S.; Wu, L. HNb$_x$Ta$_{1-x}$WO$_6$ monolayer nanosheets solid solutions: Tunable energy band structures and highly enhanced photocatalytic performances for hydrogen evolution. *Appl. Catal. B Environ.* **2017**, *203*, 798–806. [CrossRef]

61. Lozano-Sánchez, L.; Méndez-Medrano, M.; Colbeau-Justin, C.; Rodríguez-López, J.; Hernández-Uresti, D.; Obregón, S. Long-lived photoinduced charge-carriers in Er^{3+} doped CaTiO$_3$ for photocatalytic H$_2$ production under UV irradiation. *Catal. Commun.* **2016**, *84*, 36–39. [CrossRef]

62. Lucchetti, R.; Onotri, L.; Clarizia, L.; Di Natale, F.; Di Somma, I.; Andreozzi, R.; Marotta, R. Removal of nitrate and simultaneous hydrogen generation through photocatalytic reforming of glycerol over "in situ" prepared zero-valent nano copper/P25. *Appl. Catal. B Environ.* **2017**, *202*, 539–549. [CrossRef]

63. Luévano-Hipólito, E.; Torres-Martínez, L.; Sánchez-Martínez, D.; Cruz, M.A. Cu$_2$O precipitation-assisted with ultrasound and microwave radiation for photocatalytic hydrogen production. *Int. J. Hydrogen Energy* **2017**, *42*, 12997–13010. [CrossRef]

64. Luna, A.L.; Novoseltceva, E.; Louarn, E.; Beaunier, P.; Kowalska, E.; Ohtani, B.; Valenzuela, M.A.; Remita, H.; Colbeau-Justin, C. Synergetic effect of Ni and Au nanoparticles synthesized on titania particles for efficient photocatalytic hydrogen production. *Appl. Catal. B Environ.* **2016**, *191*, 18–28. [CrossRef]

65. Majeed, I.; Nadeem, M.A.; Badshah, A.; Kanodarwala, F.K.; Ali, H.; Khan, M.A.; Stride, J.A.; Nadeem, M.A. Titania supported MOF-199 derived Cu–Cu$_2$O nanoparticles: Highly efficient non-noble metal photocatalysts for hydrogen production from alcohol–water mixtures. *Catal. Sci. Technol.* **2017**, *7*, 677–686. [CrossRef]

66. Manjunath, K.; Souza, V.; Ramakrishnappa, T.; Nagaraju, G.; Scholten, J.; Dupont, J. Heterojunction CuO-TiO$_2$ nanocomposite synthesis for significant photocatalytic hydrogen production. *Mater. Res. Express* **2016**, *3*, 115904. [CrossRef]

67. Manjunath, K.; Souza, V.S.; Ganganagappa, N.; Scholten, J.D.; Teixeira, S.R.; Dupont, J.; Thippeswamy, R. Effect of the magnetic core of (MnFe)$_2$O$_3$@Ta$_2$O$_5$ nanoparticles on photocatalytic hydrogen production. *New J. Chem.* **2017**, *41*, 326–334. [CrossRef]

68. Mansingh, S.; Padhi, D.; Parida, K. Enhanced photocatalytic activity of nanostructured Fe doped CeO$_2$ for hydrogen production under visible light irradiation. *Int. J. Hydrogen Energy* **2016**, *41*, 14133–14146. [CrossRef]

69. Markovskaya, D.V.; Kozlova, E.A.; Cherepanova, S.V.; Saraev, A.A.; Gerasimov, E.Y.; Parmon, V.N. Synthesis of Pt/Zn(OH)$_2$/Cd$_{0.3}$Zn$_{0.7}$S for the Photocatalytic Hydrogen Evolution from Aqueous Solutions of Organic and Inorganic Electron Donors Under Visible Light. *Top. Catal.* **2016**, *59*, 1297–1304. [CrossRef]

70. Melián, E.P.; López, C.R.; Santiago, D.E.; Quesada-Cabrera, R.; Méndez, J.O.; Rodríguez, J.D.; Díaz, O.G. Study of the photocatalytic activity of Pt-modified commercial TiO$_2$ for hydrogen production in the presence of common organic sacrificial agents. *Appl. Catal. A Gen.* **2016**, *518*, 189–197. [CrossRef]

71. Méndez-Medrano, M.; Kowalska, E.; Lehoux, A.; Herissan, A.; Ohtani, B.; Rau, S.; Colbeau-Justin, C.; Rodríguez-López, J.; Remita, H. Surface Modification of TiO$_2$ with Au Nanoclusters for Efficient Water Treatment and Hydrogen Generation under Visible Light. *J. Phys. Chem. C* **2016**, *120*, 25010–25022. [CrossRef]

72. Meng, A.; Zhang, J.; Xu, D.; Cheng, B.; Yu, J. Enhanced photocatalytic H$_2$-production activity of anatase TiO$_2$ nanosheet by selectively depositing dual-cocatalysts on {101} and {001} facets. *Appl. Catal. B Environ.* **2016**, *198*, 286–294. [CrossRef]

73. Michal, R.; Dworniczek, E.; Caplovicova, M.; Monfort, O.; Lianos, P.; Caplovic, L.; Plesch, G. Photocatalytic properties and selective antimicrobial activity of TiO$_2$ (Eu)/CuO nanocomposite. *Appl. Surf. Sci.* **2016**, *371*, 538–546. [CrossRef]

74. Miseki, Y.; Fujiyoshi, S.; Gunji, T.; Sayama, K. Photocatalytic Z-scheme water splitting for independent H$_2$/O$_2$ production via a stepwise operation employing a vanadate redox mediator under visible light. *J. Phys. Chem. C* **2017**, *121*, 9691–9697. [CrossRef]

75. Moon, S.Y.; Naik, B.; Park, J.Y. Photocatalytic activity of metal-decorated SiO$_2$@TiO$_2$ hybrid photocatalysts under water splitting. *Korean J. Chem. Eng.* **2016**, *33*, 2325–2329. [CrossRef]

76. Nadeem, M.A.; Al-Oufi, M.; Wahab, A.K.; Anjum, D.; Idriss, H. Hydrogen Production on Ag-Pd/TiO$_2$ Bimetallic Catalysts: Is there a Combined Effect of Surface Plasmon Resonance with Schottky Mechanism on the Photo-Catalytic Activity? *ChemistrySelect* **2017**, *2*, 2754–2762. [CrossRef]

77. Narendranath, S.B.; Thekkeparambil, S.V.; George, L.; Thundiyil, S.; Devi, R.N. Photocatalytic H$_2$ evolution from water–methanol mixtures on InGaO$_3$(ZnO) m with an anisotropic layered structure modified with CuO and NiO cocatalysts. *J. Mol. Catal. A Chem.* **2016**, *415*, 82–88. [CrossRef]

78. Niu, F.; Shen, S.; Guo, L. A noble-metal-free artificial photosynthesis system with TiO$_2$ as electron relay for efficient photocatalytic hydrogen evolution. *J. Catal.* **2016**, *344*, 141–147. [CrossRef]

79. Nsib, M.F.; Saafi, S.; Rayes, A.; Moussa, N.; Houas, A. Enhanced photocatalytic performance of Ni–ZnO/Polyaniline composite for the visible-light driven hydrogen generation. *J. Energy Inst.* **2016**, *89*, 694–703. [CrossRef]

80. Obregón, S.; Lee, S.; Rodríguez-González, V. Loading effects of silver nanoparticles on hydrogen photoproduction using a Cu-TiO$_2$ photocatalyst. *Mater. Lett.* **2016**, *173*, 174–177. [CrossRef]

81. Pai, M.R.; Banerjee, A.M.; Rawool, S.A.; Singhal, A.; Nayak, C.; Ehrman, S.H.; Tripathi, A.K.; Bharadwaj, S.R. A comprehensive study on sunlight driven photocatalytic hydrogen generation using low cost nanocrystalline Cu-Ti oxides. *Sol. Energy Mater. Sol. Cells* **2016**, *154*, 104–120. [CrossRef]

82. Patra, K.K.; Gopinath, C.S. Bimetallic and Plasmonic Ag–Au on TiO$_2$ for Solar Water Splitting: An Active Nanocomposite for Entire Visible-Light-Region Absorption. *ChemCatChem* **2016**, *8*, 3294–3311. [CrossRef]

83. Pérez-Larios, A.; Hernández-Gordillo, A.; Morales-Mendoza, G.; Lartundo-Rojas, L.; Mantilla, Á.; Gómez, R. Enhancing the H$_2$ evolution from water–methanol solution using Mn^{2+}–Mn^{+3}–Mn^{4+} redox species of Mn-doped TiO$_2$ sol–gel photocatalysts. *Catal. Today* **2016**, *266*, 9–16. [CrossRef]

84. Polliotto, V.; Morra, S.; Livraghi, S.; Valetti, F.; Gilardi, G.; Giamello, E. Electron transfer and H$_2$ evolution in hybrid systems based on [FeFe]-hydrogenase anchored on modified TiO$_2$. *Int. J. Hydrogen Energy* **2016**, *41*, 10547–10556. [CrossRef]

85. Preethi, L.; Antony, R.P.; Mathews, T.; Loo, S.; Wong, L.H.; Dash, S.; Tyagi, A. Nitrogen doped anatase-rutile heterostructured nanotubes for enhanced photocatalytic hydrogen production: Promising structure for sustainable fuel production. *Int. J. Hydrogen Energy* **2016**, *41*, 5865–5877. [CrossRef]

86. Priebe, J.B.; Radnik, J.; Kreyenschulte, C.; Lennox, A.J.; Junge, H.; Beller, M.; Brückner, A. H$_2$ Generation with (Mixed) Plasmonic Cu/Au-TiO$_2$ Photocatalysts: Structure–Reactivity Relationships Assessed by in situ Spectroscopy. *ChemCatChem* **2017**, *9*, 1025–1031. [CrossRef]

87. Qiu, Y.; Ouyang, F. Fabrication of TiO$_2$ hierarchical architecture assembled by nanowires with anatase/TiO$_2$ (B) phase-junctions for efficient photocatalytic hydrogen production. *Appl. Surf. Sci.* **2017**, *403*, 691–698. [CrossRef]

88. Qiu, Y.; Ouyang, F.; Zhu, R. A facile nonaqueous route for preparing mixed-phase TiO$_2$ with high activity in photocatalytic hydrogen generation. *Int. J. Hydrogen Energy* **2017**, *42*, 11364–11371. [CrossRef]

89. Ramasami, A.K.; Ravishankar, T.N.; Nagaraju, G.; Ramakrishnappa, T.; Teixeira, S.R.; Balakrishna, R.G. Gel-combustion-synthesized ZnO nanoparticles for visible light-assisted photocatalytic hydrogen generation. *Bull. Mater. Sci.* **2017**, *40*, 345–354. [CrossRef]

90. Rather, R.A.; Singh, S.; Pal, B. Visible and direct sunlight induced H$_2$ production from water by plasmonic Ag-TiO$_2$ nanorods hybrid interface. *Sol. Energy Mater. Sol. Cells* **2017**, *160*, 463–469. [CrossRef]

91. Rather, R.A.; Singh, S.; Pal, B. A Cu^{+1}/Cu^0-TiO$_2$ mesoporous nanocomposite exhibits improved H$_2$ production from H$_2$O under direct solar irradiation. *J. Catal.* **2017**, *346*, 1–9. [CrossRef]

92. Ravishankar, T.N.; Vaz, M.D.O.; Khan, S.; Ramakrishnappa, T.; Teixeira, S.R.; Balakrishna, G.; Nagaraju, G.; Dupont, J. Ionic Liquid Assisted Hydrothermal Syntheses of TiO$_2$/CuO Nano-Composites for Enhanced Photocatalytic Hydrogen Production from Water. *ChemistrySelect* **2016**, *1*, 2199–2206. [CrossRef]

93. Ravishankar, T.N.; Vaz, M.D.O.; Ramakrishnappa, T.; Teixeira, S.R.; Dupont, J. Ionic liquid assisted hydrothermal syntheses of Au doped TiO$_2$ NPs for efficient visible-light photocatalytic hydrogen production from water, electrochemical detection and photochemical detoxification of hexavalent chromium (Cr^{6+}). *RSC Adv.* **2017**, *7*, 43233–43244. [CrossRef]

94. Reddy, N.L.; Emin, S.; Valant, M.; Shankar, M. Nanostructured Bi$_2$O$_3$@TiO$_2$ photocatalyst for enhanced hydrogen production. *Int. J. Hydrogen Energy* **2017**, *42*, 6627–6636. [CrossRef]

95. Reddy, P.A.K.; Manvitha, C.; Reddy, P.V.L.; Kim, K.-H.; Kumari, V.D. Enhanced hydrogen production activity over BiOX TiO$_2$ under solar irradiation: Improved charge transfer through bismuth oxide clusters. *J. Energy Chem.* **2017**, *26*, 390–397. [CrossRef]

96. Ren, X.; Hou, H.; Liu, Z.; Gao, F.; Zheng, J.; Wang, L.; Li, W.; Ying, P.; Yang, W.; Wu, T. Shape-Enhanced Photocatalytic Activities of Thoroughly Mesoporous ZnO Nanofibers. *Small* **2016**, *12*, 4007–4017. [CrossRef] [PubMed]

97. Rico-Oller, B.; Boudjemaa, A.; Bahruji, H.; Kebir, M.; Prashar, S.; Bachari, K.; Fajardo, M.; Gómez-Ruiz, S. Photodegradation of organic pollutants in water and green hydrogen production via methanol photoreforming of doped titanium oxide nanoparticles. *Sci. Total Environ.* **2016**, *563*, 921–932. [CrossRef] [PubMed]

98. Sadanandam, G.; Valluri, D.K.; Scurrell, M.S. Highly stabilized Ag$_2$O-loaded nano TiO$_2$ for hydrogen production from glycerol: Water mixtures under solar light irradiation. *Int. J. Hydrogen Energy* **2017**, *42*, 807–820. [CrossRef]

99. Sakata, Y.; Miyoshi, Y.; Maeda, T.; Ishikiriyama, K.; Yamazaki, Y.; Imamura, H.; Ham, Y.; Hisatomi, T.; Kubota, J.; Yamakata, A. Photocatalytic property of metal ion added SrTiO$_3$ to Overall H$_2$O splitting. *Appl. Catal. A Gen.* **2016**, *521*, 227–232. [CrossRef]

100. Salgado, S.Y.A.; Zamora, R.M.R.; Zanella, R.; Peral, J.; Malato, S.; Maldonado, M.I. Photocatalytic hydrogen production in a solar pilot plant using a Au/TiO$_2$ photo catalyst. *Int. J. Hydrogen Energy* **2016**, *41*, 11933–11940. [CrossRef]

101. Sampaio, M.J.; Oliveira, J.W.; Sombrio, C.I.; Baptista, D.L.; Teixeira, S.R.; Carabineiro, S.A.; Silva, C.G.; Faria, J.L. Photocatalytic performance of Au/ZnO nanocatalysts for hydrogen production from ethanol. *Appl. Catal. A Gen.* **2016**, *518*, 198–205. [CrossRef]

102. Siddiqi, G.; Mougel, V.; Copéret, C. Highly Active Subnanometer Au Particles Supported on TiO$_2$ for Photocatalytic Hydrogen Evolution from a Well-Defined Organogold Precursor, [Au$_5$(mesityl)$_5$]. *Inorg. Chem.* **2016**, *55*, 4026–4033. [CrossRef]

103. Song, T.; Zhang, P.; Zeng, J.; Wang, T.; Ali, A.; Zeng, H. Boosting the photocatalytic H$_2$ evolution activity of Fe$_2$O$_3$ polymorphs (α-, γ-and β-Fe$_2$O$_3$) by fullerene [C$_{60}$]-modification and dye-sensitization under visible light irradiation. *RSC Adv.* **2017**, *7*, 29184–29192. [CrossRef]

104. Sun, X.; Wang, S.; Shen, C.; Xu, X. Efficient Photocatalytic Hydrogen Production over Rh-Doped Inverse Spinel Zn$_2$TiO$_4$. *ChemCatChem* **2016**, *8*, 2289–2295. [CrossRef]

105. Taylor, S.; Mehta, M.; Barbash, D.; Samokhvalov, A. One-pot photoassisted synthesis, in situ photocatalytic testing for hydrogen generation and the mechanism of binary nitrogen and copper promoted titanium dioxide. *Photochem. Photobiol. Sci.* **2017**, *16*, 916–924. [CrossRef]

106. Tian, H.; Wang, S.; Zhang, C.; Veder, J.-P.; Pan, J.; Jaroniec, M.; Wang, L.; Liu, J. Design and synthesis of porous ZnTiO$_3$/TiO$_2$ nanocages with heterojunctions for enhanced photocatalytic H$_2$ production. *J. Mater. Chem. A* **2017**, *5*, 11615–11622. [CrossRef]

107. Tiwari, A.; Mondal, I.; Ghosh, S.; Chattopadhyay, N.; Pal, U. Fabrication of mixed phase TiO$_2$ heterojunction nanorods and their enhanced photoactivities. *Phys. Chem. Chem. Phys.* **2016**, *18*, 15260–15268. [CrossRef] [PubMed]

108. Vebber, M.; Faria, A.; Dal'Acqua, N.; Beal, L.; Fetter, G.; Machado, G.; Giovanela, M.; Crespo, J. Hydrogen production by photocatalytic water splitting using poly (allylamine hydrochloride)/poly (acrylic acid)/TiO$_2$/copper chlorophyllin self-assembled thin films. *Int. J. Hydrogen Energy* **2016**, *41*, 17995–18004. [CrossRef]

109. Wang, C.; Cai, X.; Chen, Y.; Cheng, Z.; Lin, P.; Yang, Z.; Sun, S. Improved Hydrogen Production by Glycerol Photoreforming over Ag$_2$O-TiO$_2$ Nanocomposite Mixed Oxide Synthesized by a Sol-gel Method. *Energy Procedia* **2017**, *105*, 1657–1664. [CrossRef]

110. Wang, P.; Sun, F.; Kim, J.H.; Kim, J.H.; Yang, J.; Wang, X.; Lee, J.S. Synthesis of high-purity, layered structured K$_2$Ta$_4$O$_{11}$ intermediate phase nanocrystals for photocatalytic water splitting. *Phys. Chem. Chem. Phys.* **2016**, *18*, 25831–25836. [CrossRef]

111. Wang, Q.; Jiao, D.; Wu, Y.; Guo, H.; She, H.; Li, J.; Zhong, J.; Wang, F.; Tong, J. Carbon doped solid solution Bi$_{0.5}$Dy$_{0.5}$VO$_4$ for efficient photocatalytic hydrogen evolution from water. *Int. J. Hydrogen Energy* **2016**, *41*, 16032–16039. [CrossRef]

112. Wang, M.; Shen, S.; Li, L.; Tang, Z.; Yang, J. Effects of sacrificial reagents on photocatalytic hydrogen evolution over different photocatalysts. *J. Mater. Sci.* **2017**, *52*, 5155–5164. [CrossRef]

113. Song, Y.; Wei, S.; Rong, Y.; Lu, C.; Chen, Y.; Wang, J.; Zhang, Z. Enhanced visible-light photocatalytic hydrogen evolution activity of Er^{3+}:Y$_3$Al$_5$O$_{12}$/PdS–ZnS by conduction band co-catalysts (MoO$_2$, MoS$_2$ and MoSe$_2$). *Int. J. Hydrogen Energy* **2016**, *41*, 12826–12835. [CrossRef]

114. Zhang, K.; Qian, S.; Kim, W.; Kim, J.K.; Sheng, X.; Lee, J.Y.; Park, J.H. Double 2-dimensional H$_2$-evolving catalyst tipped photocatalyst nanowires: A new avenue for high-efficiency solar to H$_2$ generation. *Nano Energy* **2017**, *34*, 481–490. [CrossRef]

115. Wu, B.-H.; Liu, W.-T.; Chen, T.-Y.; Perng, T.-P.; Huang, J.-H.; Chen, L.-J. Plasmon-enhanced photocatalytic hydrogen production on Au/TiO$_2$ hybrid nanocrystal arrays. *Nano Energy* **2016**, *27*, 412–419. [CrossRef]

116. Xie, S.; Wang, Z.; Cheng, F.; Zhang, P.; Mai, W.; Tong, Y. Ceria and ceria-based nanostructured materials for photoenergy applications. *Nano Energy* **2017**, *34*, 313–337. [CrossRef]

117. Xiong, J.; Liu, Y.; Liang, S.; Zhang, S.; Li, Y.; Wu, L. Insights into the role of Cu in promoting photocatalytic hydrogen production over ultrathin HNb$_3$O$_8$ nanosheets. *J. Catal.* **2016**, *342*, 98–104. [CrossRef]

118. Xu, D.; Hai, Y.; Zhang, X.; Zhang, S.; He, R. Bi$_2$O$_3$ cocatalyst improving photocatalytic hydrogen evolution performance of TiO$_2$. *Appl. Surf. Sci.* **2017**, *400*, 530–536. [CrossRef]

119. Liu, H.; Xu, Z.; Zhang, Z.; Ao, D. Highly efficient photocatalytic H_2 evolution from water over $CdLa_2S_4$/mesoporous gC_3N_4 hybrids under visible light irradiation. *Appl. Catal. B Environ.* **2016**, *192*, 234–241. [CrossRef]

120. Xu, L.; Sun, X.; Tu, H.; Jia, Q.; Gong, H.; Guan, J. Synchronous etching-epitaxial growth fabrication of facet-coupling $NaTaO_3/Ta_2O_5$ heterostructured nanofibers for enhanced photocatalytic hydrogen production. *Appl. Catal. B Environ.* **2016**, *184*, 309–319. [CrossRef]

121. Xu, X.; Lv, M.; Sun, X.; Liu, G. Role of surface composition upon the photocatalytic hydrogen production of Cr-doped and La/Cr-codoped $SrTiO_3$. *J. Mater. Sci.* **2016**, *51*, 6464–6473. [CrossRef]

122. Yan, X.; Xue, C.; Yang, B.; Yang, G. Novel three-dimensionally ordered macroporous Fe^{3+}-doped TiO_2 photocatalysts for H_2 production and degradation applications. *Appl. Surf. Sci.* **2017**, *394*, 248–257. [CrossRef]

123. Yang, Y.; Gao, P.; Wang, Y.; Sha, L.; Ren, X.; Zhang, J.; Yang, P.; Wu, T.; Chen, Y.; Li, X. A simple and efficient hydrogen production-storage hybrid system (Co/TiO_2) for synchronized hydrogen photogeneration with uptake. *J. Mater. Chem. A* **2017**, *5*, 9198–9203. [CrossRef]

124. Yang, Y.; Liu, G.; Irvine, J.T.; Cheng, H.M. Enhanced Photocatalytic H_2 Production in Core–Shell Engineered Rutile TiO_2. *Adv. Mater.* **2016**, *28*, 5850–5856. [CrossRef] [PubMed]

125. Yao, Z.; Zhang, Y.; He, Y.; Xia, Q.; Jiang, Z. Synthesis of hierarchical dendritic micro–nano structure $ZnFe_2O_4$ and photocatalytic activities for water splitting. *Chin. J. Chem. Eng.* **2016**, *24*, 1112–1116. [CrossRef]

126. Yu, H.; Sun, D.; Liu, J.; Fang, Y.; Li, C.C. Monodisperse mesoporous Ta_2O_5 colloidal spheres as a highly effective photocatalyst for hydrogen production. *Int. J. Hydrogen Energy* **2016**, *41*, 17225–17232. [CrossRef]

127. Yu, X.; Wei, Y.; Li, Z.; Liu, J. One-step synthesis of the single crystal Ta_2O_5 nanowires with superior hydrogen production activity. *Mater. Lett.* **2017**, *191*, 150–153. [CrossRef]

128. Ma, C.; Zhu, H.; Zhou, J.; Cui, Z.; Liu, T.; Wang, Y.; Wang, Y.; Zou, Z. Confinement effect of monolayer MoS_2 quantum dots on conjugated polyimide and promotion of solar-driven photocatalytic hydrogen generation. *Dalton Trans.* **2017**, *46*, 3877–3886. [CrossRef] [PubMed]

129. Zhang, J.; Yu, Z.; Gao, Z.; Ge, H.; Zhao, S.; Chen, C.; Chen, S.; Tong, X.; Wang, M.; Zheng, Z. Porous TiO_2 nanotubes with spatially separated platinum and CoOx cocatalysts produced by atomic layer deposition for photocatalytic hydrogen production. *Angew. Chem. Int. Ed.* **2017**, *56*, 816–820. [CrossRef]

130. Zhang, M.; Sun, R.; Li, Y.; Shi, Q.; Xie, L.; Chen, J.; Xu, X.; Shi, H.; Zhao, W. High H_2 evolution from quantum Cu (II) nanodot-doped two-dimensional ultrathin TiO_2 nanosheets with dominant exposed {001} facets for reforming glycerol with multiple electron transport pathways. *J. Phys. Chem. C* **2016**, *120*, 10746–10756. [CrossRef]

131. Zhang, Q.; Li, R.; Li, Z.; Li, A.; Wang, S.; Liang, Z.; Liao, S.; Li, C. The dependence of photocatalytic activity on the selective and nonselective deposition of noble metal cocatalysts on the facets of rutile TiO_2. *J. Catal.* **2016**, *337*, 36–44. [CrossRef]

132. Zhang, Y.J.; Zhang, L.; Kang, L.; Yang, M.Y.; Zhang, K. A new $CaWO_4$/alkali-activated blast furnace slag-based cementitious composite for production of hydrogen. *Int. J. Hydrogen Energy* **2017**, *42*, 3690–3697. [CrossRef]

133. Shen, C.-C.; Liu, Y.-N.; Zhou, X.; Guo, H.-L.; Zhao, Z.-W.; Liang, K.; Xu, A.-W. Large improvement of visible-light photocatalytic H_2-evolution based on cocatalyst-free $Zn_{0.5}Cd_{0.5}S$ synthesized through a two-step process. *Catal. Sci. Technol.* **2017**, *7*, 961–967. [CrossRef]

134. Zhu, F.; Li, C.; Ha, M.N.; Liu, Z.; Guo, Q.; Zhao, Z. Molten-salt synthesis of $Cu–SrTiO_3/TiO_2$ nanotube heterostructures for photocatalytic water splitting. *J. Mater. Sci.* **2016**, *51*, 4639–4649. [CrossRef]

135. Zhu, M.; Cai, X.; Fujitsuka, M.; Zhang, J.; Majima, T. $Au/La_2Ti_2O_7$ Nanostructures Sensitized with Black Phosphorus for Plasmon-Enhanced Photocatalytic Hydrogen Production in Visible and Near-Infrared Light. *Angew. Chem.* **2017**, *129*, 2096–2100. [CrossRef]

136. Zhu, W.; Han, D.; Niu, L.; Wu, T.; Guan, H. Z-scheme $Si/MgTiO_3$ porous heterostructures: Noble metal and sacrificial agent free photocatalytic hydrogen evolution. *Int. J. Hydrogen Energy* **2016**, *41*, 14713–14720. [CrossRef]

137. Zhu, Z.; Kao, C.-T.; Tang, B.-H.; Chang, W.-C.; Wu, R.-J. Efficient hydrogen production by photocatalytic water-splitting using Pt-doped TiO_2 hollow spheres under visible light. *Ceram. Int.* **2016**, *42*, 6749–6754. [CrossRef]

138. Zhuang, B.; Xiangqing, L.; Ge, R.; Kang, S.; Qin, L.; Li, G. Assembly and electron transfer mechanisms on visible light responsive 5, 10, 15, 20-meso-tetra (4-carboxyphenyl) porphyrin/cuprous oxide composite for photocatalytic hydrogen production. *Appl. Catal. A Gen.* **2017**, *533*, 81–89. [CrossRef]

139. Zhuang, H.; Zhang, Y.; Chu, Z.; Long, J.; An, X.; Zhang, H.; Lin, H.; Zhang, Z.; Wang, X. Synergy of metal and nonmetal dopants for visible-light photocatalysis: A case-study of Sn and N co-doped TiO_2. *Phys. Chem. Chem. Phys.* **2016**, *18*, 9636–9644. [CrossRef] [PubMed]

140. Zywitzki, D.; Jing, H.; Tüysüz, H.; Chan, C.K. High surface area, amorphous titania with reactive Ti 3+ through a photo-assisted synthesis method for photocatalytic H_2 generation. *J. Mater. Chem. A* **2017**, *5*, 10957–10967. [CrossRef]

141. Bharatvaj, J.; Preethi, V.; Kanmani, S. Hydrogen production from sulphide wastewater using Ce^{3+}–TiO_2 photocatalysis. *Int. J. Hydrogen Energy* **2018**, *43*, 3935–3945. [CrossRef]

142. Yasuda, M.; Matsumoto, T.; Yamashita, T. Sacrificial hydrogen production over TiO_2-based photocatalysts: Polyols, carboxylic acids, and saccharides. *Renew. Sustain. Energy Rev.* **2018**, *81*, 1627–1635. [CrossRef]

143. Maldonado, M.; López-Martín, A.; Colón, G.; Peral, J.; Martínez-Costa, J.; Malato, S. Solar pilot plant scale hydrogen generation by irradiation of Cu/TiO_2 composites in presence of sacrificial electron donors. *Appl. Catal. B Environ.* **2018**, *229*, 15–23. [CrossRef]

144. Hainer, A.S.; Hodgins, J.S.; Sandre, V.; Vallieres, M.; Lanterna, A.E.; Scaiano, J.C. Photocatalytic Hydrogen Generation Using Metal-Decorated TiO_2: Sacrificial Donors vs True Water Splitting. *ACS Energy Lett.* **2018**, *3*, 542–545. [CrossRef]

145. Ravi, P.; Rao, V.N.; Shankar, M.; Sathish, M. $CuOCr_2O_3$ core-shell structured co-catalysts on TiO_2 for efficient photocatalytic water splitting using direct solar light. *Int. J. Hydrogen Energy* **2018**, *43*, 3976–3987. [CrossRef]

146. Camacho, S.Y.T.; Rey, A.; Hernández-Alonso, M.D.; Llorca, J.; Medina, F.; Contreras, S. Pd/TiO_2-WO_3 photocatalysts for hydrogen generation from water-methanol mixtures. *Appl. Surf. Sci.* **2018**, *455*, 570–580. [CrossRef]

147. Huang, J.; Li, G.; Zhou, Z.; Jiang, Y.; Hu, Q.; Xue, C.; Guo, W. Efficient photocatalytic hydrogen production over Rh and Nb codoped TiO_2 nanorods. *Chem. Eng. J.* **2018**, *337*, 282–289. [CrossRef]

148. Li, Y.; Kuang, L.; Xiao, D.; Badireddy, A.R.; Hu, M.; Zhuang, S.; Wang, X.; Lee, E.S.; Marhaba, T.; Zhang, W. Hydrogen production from organic fatty acids using carbon-doped TiO_2 nanoparticles under visible light irradiation. *Int. J. Hydrogen Energy* **2018**, *43*, 4335–4346. [CrossRef]

149. Castañeda, C.; Tzompantzi, F.; Rodríguez-Rodríguez, A.; Sánchez-Dominguez, M.; Gómez, R. Improved photocatalytic hydrogen production from methanol/water solution using CuO supported on fluorinated TiO_2. *J. Chem. Technol. Biotechnol.* **2018**, *93*, 1113–1120. [CrossRef]

150. Cai, X.; Zhang, J.; Fujitsuka, M.; Majima, T. Graphitic-C_3N_4 hybridized N-doped $La_2Ti_2O_7$ two-dimensional layered composites as efficient visible-light-driven photocatalyst. *Appl. Catal. B Environ.* **2017**, *202*, 191–198. [CrossRef]

151. Cao, S.; Jiang, J.; Zhu, B.; Yu, J. Shape-dependent photocatalytic hydrogen evolution activity over a Pt nanoparticle coupled gC_3N_4 photocatalyst. *Phys. Chem. Chem. Phys.* **2016**, *18*, 19457–19463. [CrossRef]

152. Cao, S.; Yu, J. Carbon-based H_2-production photocatalytic materials. *J. Photochem. Photobiol. C Photochem. Rev.* **2016**, *27*, 72–99. [CrossRef]

153. Chang, D.W.; Baek, J.B. Nitrogen-Doped Graphene for Photocatalytic Hydrogen Generation. *Chem. Asian J.* **2016**, *11*, 1125–1137. [CrossRef]

154. Chen, F.; Yang, H.; Wang, X.; Yu, H. Facile synthesis and enhanced photocatalytic H_2-evolution performance of NiS_2-modified gC_3N_4 photocatalysts. *Chin. J. Catal.* **2017**, *38*, 296–304. [CrossRef]

155. Chen, L.; Zhou, X.; Jin, B.; Luo, J.; Xu, X.; Zhang, L.; Hong, Y. Heterojunctions in gC_3N_4/B-TiO_2 nanosheets with exposed {001} plane and enhanced visible-light photocatalytic activities. *Int. J. Hydrogen Energy* **2016**, *41*, 7292–7300. [CrossRef]

156. Chen, L.-C.; Yeh, T.-F.; Lee, Y.-L.; Teng, H. Incorporating nitrogen-doped graphene oxide dots with graphene oxide sheets for stable and effective hydrogen production through photocatalytic water decomposition. *Appl. Catal. A Gen.* **2016**, *521*, 118–124. [CrossRef]

157. Chen, T.; Quan, W.; Yu, L.; Hong, Y.; Song, C.; Fan, M.; Xiao, L.; Gu, W.; Shi, W. One-step synthesis and visible-light-driven H_2 production from water splitting of Ag quantum dots/gC_3N_4 photocatalysts. *J. Alloys Compd.* **2016**, *686*, 628–634. [CrossRef]

158. Chen, X.; Chen, H.; Guan, J.; Zhen, J.; Sun, Z.; Du, P.; Lu, Y.; Yang, S. A facile mechanochemical route to a covalently bonded graphitic carbon nitride (gC$_3$N$_4$) and fullerene hybrid toward enhanced visible light photocatalytic hydrogen production. *Nanoscale* **2017**, *9*, 5615–5623. [CrossRef]

159. Cheng, C.; Shi, J.; Hu, Y.; Guo, L. WO$_3$/g-C$_3$N$_4$ composites: One-pot preparation and enhanced photocatalytic H$_2$ production under visible-light irradiation. *Nanotechnology* **2017**, *28*, 164002. [CrossRef]

160. Cheng, R.; Zhang, L.; Fan, X.; Wang, M.; Li, M.; Shi, J. One-step construction of FeO$_x$ modified gC$_3$N$_4$ for largely enhanced visible-light photocatalytic hydrogen evolution. *Carbon* **2016**, *101*, 62–70. [CrossRef]

161. Gao, J.; Wang, Y.; Zhou, S.; Lin, W.; Kong, Y. A Facile One-Step Synthesis of Fe-Doped g-C$_3$N$_4$ Nanosheets and Their Improved Visible-Light Photocatalytic Performance. *ChemCatChem* **2017**, *9*, 1708–1715. [CrossRef]

162. Gholipour, M.R.; Béland, F.; Do, T.-O. Graphitic Carbon Nitride-Titanium Dioxide Nanocomposite for Photocatalytic Hydrogen Production under Visible Light. *Int. J. Chem. React. Eng.* **2016**, *14*, 851–858. [CrossRef]

163. Wenderich, K.; Mul, G. Methods, mechanism, and applications of photodeposition in photocatalysis: A review. *Chem. Rev.* **2016**, *116*, 14587–14619. [CrossRef]

164. Li, X.; Guo, S.; Kan, C.; Zhu, J.; Tong, T.; Ke, S.; Choy, W.C.; Wei, B. Au Multimer@ MoS$_2$ hybrid structures for efficient photocatalytical hydrogen production via strongly plasmonic coupling effect. *Nano Energy* **2016**, *30*, 549–558. [CrossRef]

165. Han, Q.; Wang, B.; Gao, J.; Qu, L. Graphitic Carbon Nitride/Nitrogen-Rich Carbon Nanofibers: Highly Efficient Photocatalytic Hydrogen Evolution without Cocatalysts. *Angew. Chem.* **2016**, *128*, 11007–11011. [CrossRef]

166. Hao, R.; Guo, S.; Wang, X.; Feng, T.; Feng, Q.; Li, M.; Jiang, B. Two-dimensional assembly structure of graphene and TiO$_2$ nanosheets from titanic acid with enhanced visible-light photocatalytic performance. *Chem. Phys. Lett.* **2016**, *653*, 190–195. [CrossRef]

167. Han, W.; Li, Z.; Li, Y.; Fan, X.; Zhang, F.; Zhang, G.; Peng, W. The Promoting Role of Different Carbon Allotropes Cocatalysts for Semiconductors in Photocatalytic Energy Generation and Pollutants Degradation. *Front. Chem.* **2017**, *5*, 84. [CrossRef]

168. He, K.; Xie, J.; Yang, Z.; Shen, R.; Fang, Y.; Ma, S.; Chen, X.; Li, X. Earth-abundant WC nanoparticles as an active noble-metal-free co-catalyst for the highly boosted photocatalytic H$_2$ production over gC$_3$N$_4$ nanosheets under visible light. *Catal. Sci. Technol.* **2017**, *7*, 1193–1202. [CrossRef]

169. Chen, Y.; He, J.; Li, J.; Mao, M.; Yan, Z.; Wang, W.; Wang, J. Hydrilla derived ZnIn$_2$S$_4$ photocatalyst with hexagonal-cubic phase junctions: A bio-inspired approach for H$_2$ evolution. *Catal. Commun.* **2016**, *87*, 1–5. [CrossRef]

170. Hong, Y.; Fang, Z.; Yin, B.; Luo, B.; Zhao, Y.; Shi, W.; Li, C. A visible-light-driven heterojunction for enhanced photocatalytic water splitting over Ta$_2$O$_5$ modified gC$_3$N$_4$ photocatalyst. *Int. J. Hydrogen Energy* **2017**, *42*, 6738–6745. [CrossRef]

171. Huang, Q.-Z.; Wang, J.-C.; Wang, P.-P.; Yao, H.-C.; Li, Z.-J. In-situ growth of mesoporous Nb$_2$O$_5$ microspheres on gC$_3$N$_4$ nanosheets for enhanced photocatalytic H$_2$ evolution under visible light irradiation. *Int. J. Hydrogen Energy* **2017**, *42*, 6683–6694. [CrossRef]

172. Huo, J.; Liu, X.; Li, X.; Qin, L.; Kang, S.-Z. An efficient photocatalytic system containing Eosin Y, 3D mesoporous graphene assembly and CuO for visible-light-driven H$_2$ evolution from water. *Int. J. Hydrogen Energy* **2017**, *42*, 15540–15550. [CrossRef]

173. Kalyani, R.; Gurunathan, K. PTh-rGO-TiO$_2$ nanocomposite for photocatalytic hydrogen production and dye degradation. *J. Photochem. Photobiol. A Chem.* **2016**, *329*, 105–112. [CrossRef]

174. Karthik, P.; Vinoth, R.; Selvam, P.; Balaraman, E.; Navaneethan, M.; Hayakawa, Y.; Neppolian, B. A visible-light active catechol–metal oxide carbonaceous polymeric material for enhanced photocatalytic activity. *J. Mater. Chem. A* **2017**, *5*, 384–396. [CrossRef]

175. Lau, V.W.H.; Klose, D.; Kasap, H.; Podjaski, F.; Pignié, M.C.; Reisner, E.; Jeschke, G.; Lotsch, B.V. Dark Photocatalysis: Storage of Solar Energy in Carbon Nitride for Time-Delayed Hydrogen Generation. *Angew. Chem. Int. Ed.* **2017**, *56*, 510–514. [CrossRef]

176. Lei, Y.; Shi, Q.; Han, C.; Wang, B.; Wu, N.; Wang, H.; Wang, Y. N-doped graphene grown on silk cocoon-derived interconnected carbon fibers for oxygen reduction reaction and photocatalytic hydrogen production. *Nano Res.* **2016**, *9*, 2498–2509. [CrossRef]

177. Litke, A.; Weber, T.; Hofmann, J.P.; Hensen, E.J. Bottlenecks limiting efficiency of photocatalytic water reduction by mixed Cd-Zn sulfides/Pt-TiO$_2$ composites. *Appl. Catal. B Environ.* **2016**, *198*, 16–24. [CrossRef]

178. Liu, B.; Su, S.; Zhou, W.; Wang, Y.; Wei, D.; Yao, L.; Ni, Y.; Cao, M.; Hu, C. Photo-reduction assisted synthesis of W-doped TiO$_2$ coupled with Au nanoparticles for highly efficient photocatalytic hydrogen evolution. *CrystEngComm* **2017**, *19*, 675–683. [CrossRef]

179. Liu, G.; Zhao, G.; Zhou, W.; Liu, Y.; Pang, H.; Zhang, H.; Hao, D.; Meng, X.; Li, P.; Kako, T. In situ bond modulation of graphitic carbon nitride to construct p–n homojunctions for enhanced photocatalytic hydrogen production. *Adv. Funct. Mater.* **2016**, *26*, 6822–6829. [CrossRef]

180. Lu, Y.; Ma, B.; Yang, Y.; Huang, E.; Ge, Z.; Zhang, T.; Zhang, S.; Li, L.; Guan, N.; Ma, Y. High activity of hot electrons from bulk 3D graphene materials for efficient photocatalytic hydrogen production. *Nano Res.* **2017**, *10*, 1662–1672. [CrossRef]

181. Ma, B.; Xu, H.; Lin, K.; Li, J.; Zhan, H.; Liu, W.; Li, C. Mo$_2$C as Non-Noble Metal Co-Catalyst in Mo$_2$C/CdS Composite for Enhanced Photocatalytic H$_2$ Evolution under Visible Light Irradiation. *ChemSusChem* **2016**, *9*, 820–824. [CrossRef]

182. Mao, Z.; Chen, J.; Yang, Y.; Wang, D.; Bie, L.; Fahlman, B.D. Novel g-C$_3$N$_4$/CoO nanocomposites with significantly enhanced visible-light photocatalytic activity for H$_2$ evolution. *ACS Appl. Mater. Interfaces* **2017**, *9*, 12427–12435. [CrossRef] [PubMed]

183. Mateo, D.; Esteve-Adell, I.; Albero, J.; Royo, J.F.S.; Primo, A.; Garcia, H. 111 oriented gold nanoplatelets on multilayer graphene as visible light photocatalyst for overall water splitting. *Nat. Commun.* **2016**, *7*, 11819. [CrossRef] [PubMed]

184. Mehta, A.; Pooja, D.; Thakur, A.; Basu, S. Enhanced photocatalytic water splitting by gold carbon dot core shell nanocatalyst under visible/sunlight. *New J. Chem.* **2017**, *41*, 4573–4581. [CrossRef]

185. Nguyen, C.C.; Vu, N.N.; Chabot, S.; Kaliaguine, S.; Do, T.O. Role of CxNy-Triazine in Photocatalysis for Efficient Hydrogen Generation and Organic Pollutant Degradation Under Solar Light Irradiation. *Sol. RRL* **2017**, *1*, 1700012. [CrossRef]

186. Xing, Z.; Zhang, J.; Cui, J.; Yin, J.; Zhao, T.; Kuang, J.; Xiu, Z.; Wan, N.; Zhou, W. Recent advances in floating TiO$_2$-based photocatalysts for environmental application. *Appl. Catal. B Environ.* **2018**, *225*, 452–467. [CrossRef]

187. Pan, Z.; Zheng, Y.; Guo, F.; Niu, P.; Wang, X. Decorating CoP and Pt nanoparticles on graphitic carbon nitride nanosheets to promote overall water splitting by conjugated polymers. *ChemSusChem* **2017**, *10*, 87–90. [CrossRef]

188. Putri, L.K.; Ng, B.-J.; Ong, W.-J.; Lee, H.W.; Chang, W.S.; Chai, S.-P. Heteroatom nitrogen-and boron-doping as a facile strategy to improve photocatalytic activity of standalone reduced graphene oxide in hydrogen evolution. *ACS Appl. Mater. Interfaces* **2017**, *9*, 4558–4569. [CrossRef]

189. Qiao, S.; Mitchell, R.W.; Coulson, B.; Jowett, D.V.; Johnson, B.R.; Brydson, R.; Isaacs, M.; Lee, A.F.; Douthwaite, R.E. Pore confinement effects and stabilization of carbon nitride oligomers in macroporous silica for photocatalytic hydrogen production. *Carbon* **2016**, *106*, 320–329. [CrossRef]

190. Qu, A.; Xu, X.; Xie, H.; Zhang, Y.; Li, Y.; Wang, J. Effects of calcining temperature on photocatalysis of gC$_3$N$_4$/TiO$_2$ composites for hydrogen evolution from water. *Mater. Res. Bull.* **2016**, *80*, 167–176. [CrossRef]

191. Raevskaya, A.E.; Panasiuk, Y.V.; Korzhak, G.V.; Stroyuk, O.L.; Kuchmiy, S.Y.; Dzhagan, V.M.; Zahn, D.R. Photocatalytic H$_2$ production from aqueous solutions of hydrazine and its derivatives in the presence of nitric-acid-activated graphitic carbon nitride. *Catal. Today* **2017**, *284*, 229–235. [CrossRef]

192. Rather, R.A.; Singh, S.; Pal, B. Core–shell morphology of Au-TiO$_2$@ graphene oxide nanocomposite exhibiting enhanced hydrogen production from water. *J. Ind. Eng. Chem.* **2016**, *37*, 288–294. [CrossRef]

193. She, X.; Liu, L.; Ji, H.; Mo, Z.; Li, Y.; Huang, L.; Du, D.; Xu, H.; Li, H. Template-free synthesis of 2D porous ultrathin nonmetal-doped gC$_3$N$_4$ nanosheets with highly efficient photocatalytic H$_2$ evolution from water under visible light. *Appl. Catal. B Environ.* **2016**, *187*, 144–153. [CrossRef]

194. Shen, L.; Xing, Z.; Zou, J.; Li, Z.; Wu, X.; Zhang, Y.; Zhu, Q.; Yang, S.; Zhou, W. Black TiO$_2$ nanobelts/g-C$_3$N$_4$ nanosheets Laminated Heterojunctions with Efficient Visible-Light-Driven Photocatalytic Performance. *Sci. Rep.* **2017**, *7*, 41978. [CrossRef] [PubMed]

195. Chen, Y.; Lu, C.; Tang, L.; Wei, S.; Song, Y.; Wang, J. Efficient photocatalytic hydrogen evolution from methanol/water splitting over Tm^{3+}, Yb^{3+}:NaYF$_4$–Er^{3+}:Y$_3$Al$_5$O$_{12}$/MoS$_2$–NaTaO$_3$ nanocomposite particles under infrared–visible light irradiation. *Sol. Energy Mater. Sol. Cells* **2016**, *149*, 128–136. [CrossRef]

196. Shin, S.R.; Park, J.H.; Kim, K.-H.; Choi, K.M.; Kang, J.K. Network of Heterogeneous Catalyst Arrays on the Nitrogen-Doped Graphene for Synergistic Solar Energy Harvesting of Hydrogen from Water. *Chem. Mater.* **2016**, *28*, 7725–7730. [CrossRef]

197. Song, K.; Xiao, F.; Zhang, L.; Yue, F.; Liang, X.; Wang, J.; Su, X. $W_{18}O_{49}$ nanowires grown on gC_3N_4 sheets with enhanced photocatalytic hydrogen evolution activity under visible light. *J. Mol. Catal. A Chem.* **2016**, *418*, 95–102. [CrossRef]

198. Sun, J.; Schmidt, B.V.; Wang, X.; Shalom, M. Self-Standing Carbon Nitride-Based Hydrogels with High Photocatalytic Activity. *ACS Appl. Mater. Interfaces* **2017**, *9*, 2029–2034. [CrossRef]

199. Sun, Q.; Wang, P.; Yu, H.; Wang, X. In situ hydrothermal synthesis and enhanced photocatalytic H_2-evolution performance of suspended rGO/gC_3N_4 photocatalysts. *J. Mol. Catal. A Chem.* **2016**, *424*, 369–376. [CrossRef]

200. Tay, Q.; Wang, X.; Zhao, X.; Hong, J.; Zhang, Q.; Xu, R.; Chen, Z. Enhanced visible light hydrogen production via a multiple heterojunction structure with defect-engineered gC_3N_4 and two-phase anatase/brookite TiO_2. *J. Catal.* **2016**, *342*, 55–62. [CrossRef]

201. Thaweesak, S.; Lyu, M.; Peerakiatkhajohn, P.; Butburee, T.; Luo, B.; Chen, H.; Wang, L. Two-dimensional $gC_3N_4/Ca_2Nb_2TaO_{10}$ nanosheet composites for efficient visible light photocatalytic hydrogen evolution. *Appl. Catal. B Environ.* **2017**, *202*, 184–190. [CrossRef]

202. Virca, C.N.; Winter, H.; Goforth, A.M.; Mackiewicz, M.R.; McCormick, T.M. Photocatalytic water reduction using a polymer coated carbon quantum dot sensitizer and a nickel nanoparticle catalyst. *Nanotechnology* **2017**, *28*, 195402. [CrossRef]

203. Zheng, D.; Zhang, G.; Hou, Y.; Wang, X. Layering MoS_2 on soft hollow gC_3N_4 nanostructures for photocatalytic hydrogen evolution. *Appl. Catal. A Gen.* **2016**, *521*, 2–8. [CrossRef]

204. Wang, J.; Feng, K.; Chen, B.; Li, Z.-J.; Meng, Q.-Y.; Zhang, L.-P.; Tung, C.-H.; Wu, L.-Z. Polymer-modified hydrophilic graphene: A promotor to photocatalytic hydrogen evolution for in situ formation of core@ shell cobalt nanocomposites. *J. Photochem. Photobiol. A Chem.* **2016**, *331*, 247–254. [CrossRef]

205. Wang, N.; Li, J.; Wu, L.; Li, X.; Shu, J. MnO_2 and carbon nanotube co-modified C_3N_4 composite catalyst for enhanced water splitting activity under visible light irradiation. *Int. J. Hydrogen Energy* **2016**, *41*, 22743–22750. [CrossRef]

206. Wang, P.; Sinev, I.; Sun, F.; Li, H.; Wang, D.; Li, Q.; Wang, X.; Marschall, R.; Wark, M. Rational fabrication of a graphitic-$C_3N_4/Sr_2KNb_5O_{15}$ nanorod composite with enhanced visible-light photoactivity for degradation of methylene blue and hydrogen production. *RSC Adv.* **2017**, *7*, 42774–42782. [CrossRef]

207. Boyjoo, Y.; Sun, H.; Liu, J.; Pareek, V.K.; Wang, S. A review on photocatalysis for air treatment: From catalyst development to reactor design. *Chem. Eng. J.* **2017**, *310*, 537–559. [CrossRef]

208. Wang, Q.; Lian, J.; Ma, Q.; Zhang, S.; He, J.; Zhong, J.; Li, J.; Huang, H.; Su, B. Preparation of carbon spheres supported CdS photocatalyst for enhancement its photocatalytic H_2 evolution. *Catal. Today* **2017**, *281*, 662–668. [CrossRef]

209. Wang, Y.; Tu, W.; Hong, J.; Zhang, W.; Xu, R. Molybdenum carbide microcrystals: Efficient and stable catalyst for photocatalytic H_2 evolution from water in the presence of dye sensitizer. *J. Mater.* **2016**, *2*, 344–349. [CrossRef]

210. Zhang, H.; Xin, C.; Wang, X.; Wang, K. Facile synthesis of Cd0.2Zn0.8 S-ethylenediamine hybrid solid solution and its improved photocatalytic performance. *Int. J. Hydrogen Energy* **2016**, *41*, 12019–12028. [CrossRef]

211. Wen, J.; Xie, J.; Shen, R.; Li, X.; Luo, X.; Zhang, H.; Zhang, A.; Bi, G. Markedly enhanced visible-light photocatalytic H_2 generation over gC_3N_4 nanosheets decorated by robust nickel phosphide ($Ni_{12}P_5$) cocatalysts. *Dalton Trans.* **2017**, *46*, 1794–1802. [CrossRef]

212. Wu, W.; Zhang, J.; Fan, W.; Li, Z.; Wang, L.; Li, X.; Wang, Y.; Wang, R.; Zheng, J.; Wu, M. Remedying defects in carbon nitride to improve both photooxidation and H_2 generation efficiencies. *ACS Catal.* **2016**, *6*, 3365–3371. [CrossRef]

213. Xie, L.; Ai, Z.; Zhang, M.; Sun, R.; Zhao, W. Enhanced Hydrogen Evolution in the Presence of Plasmonic Au-Photo-Sensitized g-C_3N_4 with an Extended Absorption Spectrum from 460 to 640 nm. *PLoS ONE* **2016**, *11*, e0161397. [CrossRef]

214. Xing, W.; Chen, G.; Li, C.; Sun, J.; Han, Z.; Zhou, Y.; Hu, Y.; Meng, Q. Construction of Large-Scale Ultrathin Graphitic Carbon Nitride Nanosheets by a Hydrogen-Bond-Assisted Strategy for Improved Photocatalytic Hydrogen Production and Ciprofloxacin Degradation Activity. *ChemCatChem* **2016**, *8*, 2838–2845. [CrossRef]

215. Feng, W.; Zhang, L.; Zhang, Y.; Yang, Y.; Fang, Z.; Wang, B.; Zhang, S.; Liu, P. Near-infrared-activated NaYF$_4$:Yb^{3+}, Er^{3+}/Au/CdS for H$_2$ production via photoreforming of bio-ethanol: Plasmonic Au as light nanoantenna, energy relay, electron sink and co-catalyst. *J. Mater. Chem. A* **2017**, *5*, 10311–10320. [CrossRef]

216. Yan, M.; Hua, Y.; Zhu, F.; Sun, L.; Gu, W.; Shi, W. Constructing nitrogen doped graphene quantum dots-ZnNb 2 O 6/gC$_3$N$_4$ catalysts for hydrogen production under visible light. *Appl. Catal. B Environ.* **2017**, *206*, 531–537. [CrossRef]

217. Yang, C.; Zhang, X.; Qin, J.; Shen, X.; Yu, R.; Ma, M.; Liu, R. Porous carbon-doped TiO$_2$ on TiC nanostructures for enhanced photocatalytic hydrogen production under visible light. *J. Catal.* **2017**, *347*, 36–44. [CrossRef]

218. Yang, L.; Huang, J.; Shi, L.; Cao, L.; Yu, Q.; Jie, Y.; Fei, J.; Ouyang, H.; Ye, J. A surface modification resultant thermally oxidized porous g-C$_3$N$_4$ with enhanced photocatalytic hydrogen production. *Appl. Catal. B Environ.* **2017**, *204*, 335–345. [CrossRef]

219. Ye, P.; Liu, X.; Iocozzia, J.; Yuan, Y.; Gu, L.; Xu, G.; Lin, Z. A highly stable non-noble metal Ni$_2$P co-catalyst for increased H$_2$ generation by gC$_3$N$_4$ under visible light irradiation. *J. Mater. Chem. A* **2017**, *5*, 8493–8498. [CrossRef]

220. Di, T.; Zhu, B.; Zhang, J.; Cheng, B.; Yu, J. Enhanced photocatalytic H$_2$ production on CdS nanorod using cobalt-phosphate as oxidation cocatalyst. *Appl. Surf. Sci.* **2016**, *389*, 775–782. [CrossRef]

221. Yue, X.; Yi, S.; Wang, R.; Zhang, Z.; Qiu, S. A novel architecture of dandelion-like Mo$_2$C/TiO$_2$ heterojunction photocatalysts towards high-performance photocatalytic hydrogen production from water splitting. *J. Mater. Chem. A* **2017**, *5*, 10591–10598. [CrossRef]

222. Zeng, Y.; Wang, Y.; Chen, J.; Jiang, Y.; Kiani, M.; Li, B.; Wang, R. Fabrication of high-activity hybrid NiTiO$_3$/gC$_3$N$_4$ heterostructured photocatalysts for water splitting to enhanced hydrogen production. *Ceram. Int.* **2016**, *42*, 12297–12305. [CrossRef]

223. Zeng, Z.; Li, K.; Wei, K.; Dai, Y.; Yan, L.; Guo, H.; Luo, X. Fabrication of porous gC$_3$N$_4$ and supported porous gC$_3$N$_4$ by a simple precursor pretreatment strategy and their efficient visible-light photocatalytic activity. *Chin. J. Catal.* **2017**, *38*, 498–507. [CrossRef]

224. Zha, D.-W.; Li, L.-F.; Pan, Y.-X.; He, J.-B. Coconut shell carbon nanosheets facilitating electron transfer for highly efficient visible-light-driven photocatalytic hydrogen production from water. *Int. J. Hydrogen Energy* **2016**, *41*, 17370–17379. [CrossRef]

225. Ge, M.; Li, Q.; Cao, C.; Huang, J.; Li, S.; Zhang, S.; Chen, Z.; Zhang, K.; Al-Deyab, S.S.; Lai, Y. One-dimensional TiO$_2$ Nanotube Photocatalysts for Solar Water Splitting. *Adv. Sci.* **2017**, *4*, 1600152. [CrossRef]

226. Zhang, P.; Song, T.; Wang, T.; Zeng, H. In-situ synthesis of Cu nanoparticles hybridized with carbon quantum dots as a broad spectrum photocatalyst for improvement of photocatalytic H$_2$ evolution. *Appl. Catal. B Environ.* **2017**, *206*, 328–335. [CrossRef]

227. Zhou, X.; Sun, H.; Zhang, H.; Tu, W. One-pot hydrothermal synthesis of CdS/NiS photocatalysts for high H$_2$ evolution from water under visible light. *Int. J. Hydrogen Energy* **2017**, *42*, 11199–11205. [CrossRef]

228. Zhao, W.; Xie, L.; Zhang, M.; Ai, Z.; Xi, H.; Li, Y.; Shi, Q.; Chen, J. Enhanced photocatalytic activity of all-solid-state gC$_3$N$_4$/Au/P$_{25}$ Z-scheme system for visible-light-driven H$_2$ evolution. *Int. J. Hydrogen Energy* **2016**, *41*, 6277–6287. [CrossRef]

229. Zou, J.-P.; Wang, L.-C.; Luo, J.; Nie, Y.-C.; Xing, Q.-J.; Luo, X.-B.; Du, H.-M.; Luo, S.-L.; Suib, S.L. Synthesis and efficient visible light photocatalytic H$_2$ evolution of a metal-free gC$_3$N$_4$/graphene quantum dots hybrid photocatalyst. *Appl. Catal. B Environ.* **2016**, *193*, 103–109. [CrossRef]

230. Chu, J.; Han, X.; Yu, Z.; Du, Y.; Song, B.; Xu, P. Highly Efficient Visible-Light-Driven Photocatalytic Hydrogen Production on CdS/Cu$_7$S$_4$/g-C$_3$N$_4$ Ternary Heterostructures. *ACS Appl. Mater. Interfaces* **2018**, *10*, 20404–20411. [CrossRef]

231. Hua, S.; Qu, D.; An, L.; Jiang, W.; Wen, Y.; Wang, X.; Sun, Z. Highly efficient p-type Cu$_3$P/n-type g-C$_3$N$_4$ photocatalyst through Z-scheme charge transfer route. *Appl. Catal. B Environ.* **2019**, *240*, 253–261. [CrossRef]

232. Wang, P.; Guan, Z.; Li, Q.; Yang, J. Efficient visible-light-driven photocatalytic hydrogen production from water by using Eosin Y-sensitized novel g-C$_3$N$_4$/Pt/GO composites. *J. Mater. Sci.* **2018**, *53*, 774–786. [CrossRef]

233. Liu, J.; Jia, Q.; Long, J.; Wang, X.; Gao, Z.; Gu, Q. Amorphous NiO as co-catalyst for enhanced visible-light-driven hydrogen generation over g-C$_3$N$_4$ photocatalyst. *Appl. Catal. B Environ.* **2018**, *222*, 35–43. [CrossRef]

234. Fu, J.; Xu, Q.; Low, J.; Jiang, C.; Yu, J. Ultrathin 2D/2D WO_3/g-C_3N_4 step-scheme H_2-production photocatalyst. *Appl. Catal. B Environ.* **2019**, *243*, 556–565. [CrossRef]

235. de Castro, S.; da Silva, A.F.; Felix, J.F.; Piton, M.R.; Galeti, H.V.A.; Rodrigues, A.D.G.; Gobato, Y.G.; Al Saqri, N.; Henini, M.; Albadri, A.M.; et al. Effect of growth techniques on the structural, optical and electrical properties of indium doped TiO_2 thin films. *J. Alloys Compd.* **2018**, *766*, 194–203.

236. Yu, Y.; Yan, W.; Wang, X.; Li, P.; Gao, W.; Zou, H.; Wu, S.; Ding, K. Surface Engineering for Extremely Enhanced Charge Separation and Photocatalytic Hydrogen Evolution on g-C_3N_4. *Adv. Mater.* **2018**, *30*, 1705060. [CrossRef] [PubMed]

237. Ma, B.; Li, D.; Wang, X.; Lin, K. Molybdenum-Based Co-catalysts in Photocatalytic Hydrogen Production: Categories, Structures, and Roles. *ChemSusChem* **2018**, *11*, 3871–3881. [CrossRef]

238. Acar, C.; Dincer, I.; Naterer, G.F. Review of photocatalytic water-splitting methods for sustainable hydrogen production. *Int. J. Energy Res.* **2016**, *40*, 1449–1473. [CrossRef]

239. Alkaim, A.F.; Kandiel, T.A.; Dillert, R.; Bahnemann, D.W. Photocatalytic hydrogen production from biomass-derived compounds: A case study of citric acid. *Environ. Technol.* **2016**, *37*, 2687–2693. [CrossRef]

240. Almahdi, M.; Dincer, I.; Rosen, M. Analysis and assessment of methanol production by integration of carbon capture and photocatalytic hydrogen production. *Int. J. Greenh. Gas Control* **2016**, *51*, 56–70. [CrossRef]

241. Cai, Q.; Hu, Z.; Zhang, Q.; Li, B.; Shen, Z. Fullerene (C_{60})/CdS nanocomposite with enhanced photocatalytic activity and stability. *Appl. Surf. Sci.* **2017**, *403*, 151–158. [CrossRef]

242. Cai, X.; Li, G.; Yang, Y.; Zhang, C.; Yang, X. Cobalt thiolate complexes catalyst in noble-metal-free system for photocatalytic hydrogen production. *Russ. J. Appl. Chem.* **2016**, *89*, 1506–1511. [CrossRef]

243. Chang, C.-J.; Chu, K.-W. ZnS/polyaniline composites with improved dispersing stability and high photocatalytic hydrogen production activity. *Int. J. Hydrogen Energy* **2016**, *41*, 21764–21773. [CrossRef]

244. Chang, C.-J.; Lee, Z.; Chu, K.-W.; Wei, Y.-H. $CoFe_2O_4$@ ZnS core–shell spheres as magnetically recyclable photocatalysts for hydrogen production. *J. Taiwan Ins. Chem. Eng.* **2016**, *66*, 386–393. [CrossRef]

245. Chang, K.; Hai, X.; Ye, J. Transition Metal Disulfides as Noble-Metal-Alternative Co-Catalysts for Solar Hydrogen Production. *Adv. Energy Mater.* **2016**, *6*, 1502555. [CrossRef]

246. Chen, H.; Jiang, D.; Sun, Z.; Irfan, R.M.; Zhang, L.; Du, P. Cobalt nitride as an efficient cocatalyst on CdS nanorods for enhanced photocatalytic hydrogen production in water. *Catal. Sci. Technol.* **2017**, *7*, 1515–1522. [CrossRef]

247. Chen, T.; Song, C.; Fan, M.; Hong, Y.; Hu, B.; Yu, L.; Shi, W. In-situ fabrication of CuS/gC_3N_4 nanocomposites with enhanced photocatalytic H_2-production activity via photoinduced interfacial charge transfer. *Int. J. Hydrogen Energy* **2017**, *42*, 12210–12219. [CrossRef]

248. Chen, Y.; Tian, G.; Zhou, W.; Xiao, Y.; Wang, J.; Zhang, X.; Fu, H. Enhanced photogenerated carrier separation in CdS quantum dot sensitized $ZnFe_2O_4$/$ZnIn_2S_4$ nanosheet stereoscopic films for exceptional visible light photocatalytic H_2 evolution performance. *Nanoscale* **2017**, *9*, 5912–5921. [CrossRef]

249. Chu, J.; Han, X.; Yu, Z.; Du, Y.; Song, B.; Xu, P. Fabrication of H-TiO_2/CdS/Cu_2-xS Ternary Heterostructures for Enhanced Photocatalytic Hydrogen Production. *ChemistrySelect* **2017**, *2*, 2684–2689. [CrossRef]

250. Dong, M.; Zhou, P.; Jiang, C.; Cheng, B.; Yu, J. First-principles investigation of Cu-doped ZnS with enhanced photocatalytic hydrogen production activity. *Chem. Phys. Lett.* **2017**, *668*, 1–6. [CrossRef]

251. Du, R.; Zhang, Y.; Li, B.; Yu, X.; Liu, H.; An, X.; Qu, J. Biomolecule-assisted synthesis of defect-mediated $Cd_{1-x}Zn_xS$/MoS_2/graphene hollow spheres for highly efficient hydrogen evolution. *Phys. Chem. Chem. Phys.* **2016**, *18*, 16208–16215. [CrossRef]

252. Fang, X.; Song, J.; Shi, H.; Kang, S.; Li, Y.; Sun, G.; Cui, L. Enhanced efficiency and stability of $Co_{0.5}Cd_{0.5}S$/gC_3N_4 composite photo-catalysts for hydrogen evolution from water under visible light irradiation. *Int. J. Hydrogen Energy* **2017**, *42*, 5741–5748. [CrossRef]

253. Feng, W.; Fang, Z.; Wang, B.; Zhang, L.; Zhang, Y.; Yang, Y.; Huang, M.; Weng, S.; Liu, P. Grain boundary engineering in organic–inorganic hybrid semiconductor ZnS (en) 0.5 for visible-light photocatalytic hydrogen production. *J. Mater. Chem. A* **2017**, *5*, 1387–1393. [CrossRef]

254. García-Mendoza, C.; Oros-Ruiz, S.; Hernández-Gordillo, A.; López, R.; Jácome-Acatitla, G.; Calderón, H.A.; Gómez, R. Suitable preparation of Bi_2S_3 nanorods–TiO_2 heterojunction semiconductors with improved photocatalytic hydrogen production from water/methanol decomposition. *J. Chem. Technol. Biotechnol.* **2016**, *91*, 2198–2204. [CrossRef]

255. Gu, Q.; Sun, H.; Xie, Z.; Gao, Z.; Xue, C. MoS_2-coated microspheres of self-sensitized carbon nitride for efficient photocatalytic hydrogen generation under visible light irradiation. *Appl. Surf. Sci.* **2017**, *396*, 1808–1815. [CrossRef]

256. Guo, H.-L.; Du, H.; Jiang, Y.-F.; Jiang, N.; Shen, C.-C.; Zhou, X.; Liu, Y.-N.; Xu, A.-W. Artificial Photosynthetic Z-scheme Photocatalyst for Hydrogen Evolution with High Quantum Efficiency. *J. Phys. Chem. C* **2016**, *121*, 107–114. [CrossRef]

257. Guo, X.; Chen, Y.; Qin, Z.; Wang, M.; Guo, L. One-step hydrothermal synthesis of $Zn_xCd_{1-x}S/ZnO$ heterostructures for efficient photocatalytic hydrogen production. *Int. J. Hydrogen Energy* **2016**, *41*, 15208–15217. [CrossRef]

258. Ha, E.; Liu, W.; Wang, L.; Man, H.-W.; Hu, L.; Tsang, S.C.E.; Chan, C.T.-L.; Kwok, W.-M.; Lee, L.Y.S.; Wong, K.-Y. Cu_2ZnSnS_4/MoS_2-Reduced Graphene Oxide Heterostructure: Nanoscale Interfacial Contact and Enhanced Photocatalytic Hydrogen Generation. *Sci. Rep.* **2017**, *7*, 39411. [CrossRef]

259. Hai, X.; Zhou, W.; Chang, K.; Pang, H.; Liu, H.; Shi, L.; Ichihara, F.; Ye, J. Engineering the crystallinity of MoS 2 monolayers for highly efficient solar hydrogen production. *J. Mater. Chem. A* **2017**, *5*, 8591–8598. [CrossRef]

260. Hong, S.; Kumar, D.P.; Reddy, D.A.; Choi, J.; Kim, T.K. Excellent photocatalytic hydrogen production over CdS nanorods via using noble metal-free copper molybdenum sulfide (Cu_2MoS_4) nanosheets as co-catalysts. *Appl. Surf. Sci.* **2017**, *396*, 421–429. [CrossRef]

261. Hu, P.; Ngaw, C.K.; Yuan, Y.; Bassi, P.S.; Loo, S.C.J.; Tan, T.T.Y. Bandgap engineering of ternary sulfide nanocrystals by solution proton alloying for efficient photocatalytic H_2 evolution. *Nano Energy* **2016**, *26*, 577–585. [CrossRef]

262. Hu, S.; Zhu, M. Enhanced Solar Hydrogen Generation by a Heterojunction of Perovskite-type $La_2Ti_2O_7$ Nanosheets Doped with CdS Quantum Dots. *ChemPlusChem* **2016**, *81*, 1202–1208. [CrossRef]

263. Huang, E.; Yao, X.; Wang, W.; Wu, G.; Guan, N.; Li, L. SnS_2 Nanoplates with Specific Facets Exposed for Enhanced Visible-Light-Driven Photocatalysis. *ChemPhotoChem* **2017**, *1*, 60–69. [CrossRef]

264. Huang, H.; Dai, B.; Wang, W.; Lu, C.; Kou, J.; Ni, Y.; Wang, L.; Xu, Z. Oriented Built-in Electric Field Introduced by Surface Gradient Diffusion Doping for Enhanced Photocatalytic H_2 Evolution in CdS Nanorods. *Nano Lett.* **2017**, *17*, 3803–3808. [CrossRef]

265. Huang, T.; Chen, W.; Liu, T.-Y.; Hao, Q.-L.; Liu, X.-H. Hybrid of AgInZnS and MoS_2 as efficient visible-light driven photocatalyst for hydrogen production. *Int. J. Hydrogen Energy* **2017**, *42*, 12254–12261. [CrossRef]

266. Huang, T.; Chen, W.; Liu, T.-Y.; Hao, Q.-L.; Liu, X.-H. $ZnIn_2S_4$ hybrid with MoS_2: A non-noble metal photocatalyst with efficient photocatalytic activity for hydrogen evolution. *Powder Technol.* **2017**, *315*, 157–162. [CrossRef]

267. Irfan, R.M.; Jiang, D.; Sun, Z.; Lu, D.; Du, P. Enhanced photocatalytic H_2 production on CdS nanorods with simple molecular bidentate cobalt complexes as cocatalysts under visible light. *Dalton Trans.* **2016**, *45*, 12897–12905. [CrossRef]

268. Jiang, F.; Pan, B.; You, D.; Zhou, Y.; Wang, X.; Su, W. Visible light photocatalytic H_2-production activity of epitaxial Cu_2ZnSnS_4/ZnS heterojunction. *Catal. Commun.* **2016**, *85*, 39–43. [CrossRef]

269. Jiang, Z.; Liu, J.; Gao, M.; Fan, X.; Zhang, L.; Zhang, J. Assembling Polyoxo-Titanium Clusters and CdS Nanoparticles to a Porous Matrix for Efficient and Tunable H_2-Evolution Activities with Visible Light. *Adv. Mater.* **2017**, *29*, 1603369. [CrossRef]

270. Jo, W.-K.; Selvam, N.C.S. Fabrication of photostable ternary $CdS/MoS_2/MWCNTs$ hybrid photocatalysts with enhanced H_2 generation activity. *Appl. Catal. A Gen.* **2016**, *525*, 9–22. [CrossRef]

271. Kandiel, T.A.; Takanabe, K. Solvent-induced deposition of Cu–Ga–In–S nanocrystals onto a titanium dioxide surface for visible-light-driven photocatalytic hydrogen production. *Appl. Catal. B Environ.* **2016**, *184*, 264–269. [CrossRef]

272. Kaur, M.; Nagaraja, C. Template-Free Synthesis of $Zn_{1-x}Cd_xS$ Nanocrystals with Tunable Band Structure for Efficient Water Splitting and Reduction of Nitroaromatics in Water. *ACS Sustain. Chem. Eng.* **2017**, *5*, 4293–4303. [CrossRef]

273. Kim, Y.G.; Jo, W.-K. Photodeposited-metal/CdS/ZnO heterostructures for solar photocatalytic hydrogen production under different conditions. *Int. J. Hydrogen Energy* **2017**, *42*, 11356–11363. [CrossRef]

274. Kim, Y.K.; Lim, S.K.; Park, H.; Hoffmann, M.R.; Kim, S. Trilayer CdS/carbon nanofiber (CNF) mat/Pt-TiO_2 composite structures for solar hydrogen production: Effects of CNF mat thickness. *Appl. Catal. B Environ.* **2016**, *196*, 216–222. [CrossRef]

275. Kimi, M.; Yuliati, L.; Shamsuddin, M. Preparation and characterization of In and Cu co-doped ZnS photocatalysts for hydrogen production under visible light irradiation. *J. Energy Chem.* **2016**, *25*, 512–516. [CrossRef]

276. Kong, Z.; Yuan, Y.-J.; Chen, D.; Fang, G.; Yang, Y.; Yang, S.; Cao, D. Noble-metal-free MoS_2 nanosheet modified-$InVO_4$ heterostructures for enhanced visible-light-driven photocatalytic H_2 production. *Dalton Trans.* **2017**, *46*, 2072–2076. [CrossRef] [PubMed]

277. Kumar, D.P.; Hong, S.; Reddy, D.A.; Kim, T.K. Ultrathin MoS_2 layers anchored exfoliated reduced graphene oxide nanosheet hybrid as a highly efficient cocatalyst for CdS nanorods towards enhanced photocatalytic hydrogen production. *Appl. Catal. B Environ.* **2017**, *212*, 7–14. [CrossRef]

278. Leo, I.M.; Soto, E.; Vaquero, F.; Mota, N.; Navarro, R.; Fierro, J. Influence of the reduction of graphene oxide (rGO) on the structure and photoactivity of CdS-rGO hybrid systems. *Int. J. Hydrogen Energy* **2017**, *42*, 13691–13703.

279. Li, M.; Zhang, L.; Fan, X.; Wu, M.; Du, Y.; Wang, M.; Kong, Q.; Zhang, L.; Shi, J. Dual synergetic effects in MoS_2/pyridine-modified gC_3N_4 composite for highly active and stable photocatalytic hydrogen evolution under visible light. *Appl. Catal. B Environ.* **2016**, *190*, 36–43. [CrossRef]

280. Li, X.; Liu, H.; Liu, S.; Zhang, J.; Chen, W.; Huang, C.; Mao, L. Effect of Pt–Pd hybrid nano-particle on CdS's activity for water splitting under visible light. *Int. J. Hydrogen Energy* **2016**, *41*, 23015–23021. [CrossRef]

281. Li, Y.; Hou, Y.; Fu, Q.; Peng, S.; Hu, Y.H. Oriented growth of $ZnIn_2S_4$/$In(OH)_3$ heterojunction by a facile hydrothermal transformation for efficient photocatalytic H_2 production. *Appl. Catal. B Environ.* **2017**, *206*, 726–733. [CrossRef]

282. Li, Y.; Jin, R.; Xing, Y.; Li, J.; Song, S.; Liu, X.; Li, M.; Jin, R. Macroscopic Foam-Like Holey Ultrathin g-C_3N_4 Nanosheets for Drastic Improvement of Visible-Light Photocatalytic Activity. *Adv. Energy Mater.* **2016**, *6*, 1601273. [CrossRef]

283. Li, Z.; Chen, X.; Shangguan, W.; Su, Y.; Liu, Y.; Dong, X.; Sharma, P.; Zhang, Y. Prickly Ni_3S_2 nanowires modified CdS nanoparticles for highly enhanced visible-light photocatalytic H_2 production. *Int. J. Hydrogen Energy* **2017**, *42*, 6618–6626. [CrossRef]

284. Lin, H.; Li, Y.; Li, H.; Wang, X. Multi-node CdS hetero-nanowires grown with defect-rich oxygen-doped MoS_2 ultrathin nanosheets for efficient visible-light photocatalytic H_2 evolution. *Nano Res.* **2017**, *10*, 1377–1392. [CrossRef]

285. Liu, H.; Xu, Z.; Zhang, Z.; Ao, D. Novel visible-light driven $Mn_{0.8}Cd_{0.2}S$/gC_3N_4 composites: Preparation and efficient photocatalytic hydrogen production from water without noble metals. *Appl. Catal. A Gen.* **2016**, *518*, 150–157. [CrossRef]

286. Liu, M.; Chen, Y.; Su, J.; Shi, J.; Wang, X.; Guo, L. Photocatalytic hydrogen production using twinned nanocrystals and an unanchored NiSx co-catalyst. *Nat. Energy* **2016**, *1*, 16151. [CrossRef]

287. Liu, X.; Xing, Z.; Zhang, Y.; Li, Z.; Wu, X.; Tan, S.; Yu, X.; Zhu, Q.; Zhou, W. Fabrication of 3D flower-like black N-TiO_2-x@ MoS_2 for unprecedented-high visible-light-driven photocatalytic performance. *Appl. Catal. B Environ.* **2017**, *201*, 119–127. [CrossRef]

288. Liu, Y.; Tang, C. Enhancement of photocatalytic H_2 evolution over TiO_2 nano-sheet films by surface loading NiS nanoparticles. *Russ. J. Phys. Chem. A* **2016**, *90*, 1042–1048. [CrossRef]

289. Lu, D.; Wang, H.; Zhao, X.; Kondamareddy, K.K.; Ding, J.; Li, C.; Fang, P. Highly efficient visible-light-induced photoactivity of Z-scheme g-C_3N_4/Ag/MoS_2 ternary photocatalysts for organic pollutant degradation and production of hydrogen. *ACS Sustain. Chem. Eng.* **2017**, *5*, 1436–1445. [CrossRef]

290. Ma, L.; Chen, K.; Nan, F.; Wang, J.H.; Yang, D.J.; Zhou, L.; Wang, Q.Q. Improved Hydrogen Production of Au–Pt–CdS Hetero-Nanostructures by Efficient Plasmon-Induced Multipathway Electron Transfer. *Adv. Funct. Mater.* **2016**, *26*, 6076–6083. [CrossRef]

291. Ma, X.; Li, J.; An, C.; Feng, J.; Chi, Y.; Liu, J.; Zhang, J.; Sun, Y. Ultrathin Co (Ni)-doped MoS- nanosheets as catalytic promoters enabling efficient solar hydrogen production. *Nano Res.* **2016**, *9*, 2284–2293. [CrossRef]

292. Majeed, I.; Nadeem, M.A.; Hussain, E.; Badshah, A.; Gilani, R.; Nadeem, M.A. Effect of deposition method on metal loading and photocatalytic activity of Au/CdS for hydrogen production in water electrolyte mixture. *Int. J. Hydrogen Energy* **2017**, *42*, 3006–3018. [CrossRef]

293. Malekshoar, G.; Ray, A.K. In-situ grown molybdenum sulfide on TiO_2 for dye-sensitized solar photocatalytic hydrogen generation. *Chem. Eng. Sci.* **2016**, *152*, 35–44. [CrossRef]

294. Mancipe, S.; Tzompantzi, F.; Gómez, R. Synthesis of CdS/MgAl layered double hydroxides for hydrogen production from methanol-water decomposition. *Appl. Clay Sci.* **2017**, *136*, 67–74. [CrossRef]

295. Manjunath, K.; Souza, V.; Nagaraju, G.; Santos, J.M.L.; Dupont, J.; Ramakrishnappa, T. Superior activity of the CuS–TiO$_2$/Pt hybrid nanostructure towards visible light induced hydrogen production. *New J. Chem.* **2016**, *40*, 10172–10180. [CrossRef]

296. Mei, Z.; Zhang, M.; Schneider, J.; Wang, W.; Zhang, N.; Su, Y.; Chen, B.; Wang, S.; Rogach, A.L.; Pan, F. Hexagonal Zn$_{1-x}$Cd$_x$S (0.2 ≤ x ≤ 1) solid solution photocatalysts for H$_2$ generation from water. *Catal. Sci. Technol.* **2017**, *7*, 982–987. [CrossRef]

297. Nandy, S.; Goto, Y.; Hisatomi, T.; Moriya, Y.; Minegishi, T.; Katayama, M.; Domen, K. Synthesis and Photocatalytic Activity of La$_5$Ti$_2$Cu (S$_{1-x}$Se$_x$)$_5$O$_7$ Solid Solutions for H$_2$ Production under Visible Light Irradiation. *ChemPhotoChem* **2017**, *1*, 265–272. [CrossRef]

298. Núñez, J.; Fresno, F.; Collado, L.; Jana, P.; Coronado, J.M.; Serrano, D.P.; Víctor, A. Photocatalytic H$_2$ production from aqueous methanol solutions using metal-co-catalysed Zn$_2$SnO$_4$ nanostructures. *Appl. Catal. B Environ.* **2016**, *191*, 106–115. [CrossRef]

299. Oros-Ruiz, S.; Hernández-Gordillo, A.; García-Mendoza, C.; Rodríguez-Rodríguez, A.A.; Gomez, R. Comparative activity of CdS nanofibers superficially modified by Au, Cu, and Ni nanoparticles as co-catalysts for photocatalytic hydrogen production under visible light. *J. Chem. Technol. Biotechnol.* **2016**, *91*, 2205–2210. [CrossRef]

300. Park, H.; Ou, H.-H.; Kim, M.; Kang, U.; Han, D.S.; Hoffmann, M.R. Photocatalytic H$_2$ production on trititanate nanotubes coupled with CdS and platinum nanoparticles under visible light: Revisiting H$_2$ production and material durability. *Faraday Discuss.* **2017**, *198*, 419–431. [CrossRef]

301. Qiu, F.; Han, Z.; Peterson, J.J.; Odoi, M.Y.; Sowers, K.L.; Krauss, T.D. Photocatalytic hydrogen generation by CdSe/CdS nanoparticles. *Nano Lett.* **2016**, *16*, 5347–5352. [CrossRef] [PubMed]

302. Rahman, M.; Davey, K.; Qiao, S.Z. Counteracting Blueshift Optical Absorption and Maximizing Photon Harvest in Carbon Nitride Nanosheets Photocatalyst. *Small* **2017**, *13*, 1700376. [CrossRef]

303. Rahmawati, F.; Yuliati, L.; Alaih, I.S.; Putri, F.R. Carbon rod of zinc-carbon primary battery waste as a substrate for CdS and TiO$_2$ photocatalyst layer for visible light driven photocatalytic hydrogen production. *J. Environ. Chem. Eng.* **2017**, *5*, 2251–2258. [CrossRef]

304. Ran, J.; Gao, G.; Li, F.-T.; Ma, T.-Y.; Du, A.; Qiao, S.-Z. Ti$_3$C$_2$ MXene co-catalyst on metal sulfide photo-absorbers for enhanced visible-light photocatalytic hydrogen production. *Nat. Commun.* **2017**, *8*, 13907. [CrossRef]

305. Rao, H.; Yu, W.-Q.; Zheng, H.-Q.; Bonin, J.; Fan, Y.-T.; Hou, H.-W. Highly efficient photocatalytic hydrogen evolution from nickel quinolinethiolate complexes under visible light irradiation. *J. Power Sources* **2016**, *324*, 253–260. [CrossRef]

306. Reddy, D.A.; Park, H.; Hong, S.; Kumar, D.P.; Kim, T.K. Hydrazine-assisted formation of ultrathin MoS$_2$ nanosheets for enhancing their co-catalytic activity in photocatalytic hydrogen evolution. *J. Mater. Chem. A* **2017**, *5*, 6981–6991. [CrossRef]

307. Reddy, D.A.; Park, H.; Ma, R.; Kumar, D.P.; Lim, M.; Kim, T.K. Heterostructured WS$_2$-MoS$_2$ Ultrathin Nanosheets Integrated on CdS Nanorods to Promote Charge Separation and Migration and Improve Solar-Driven Photocatalytic Hydrogen Evolution. *ChemSusChem* **2017**, *10*, 1563–1570. [CrossRef] [PubMed]

308. Shi, Z.; Dong, X.; Dang, H. Facile fabrication of novel red phosphorus-CdS composite photocatalysts for H$_2$ evolution under visible light irradiation. *Int. J. Hydrogen Energy* **2016**, *41*, 5908–5915. [CrossRef]

309. Sola, A.; Homs, N.; de la Piscina, P.R. Photocatalytic H$_2$ production from ethanol (aq) solutions: The effect of intermediate products. *Int. J. Hydrogen Energy* **2016**, *41*, 19629–19636. [CrossRef]

310. Souza, E.A.; Silva, L.A. Energy recovery from tannery sludge wastewaters through photocatalytic hydrogen production. *J. Environ. Chem. Eng.* **2016**, *4*, 2114–2120. [CrossRef]

311. Su, J.; Zhang, T.; Li, Y.; Chen, Y.; Liu, M. Photocatalytic activities of copper doped cadmium sulfide microspheres prepared by a facile ultrasonic spray-pyrolysis method. *Molecules* **2016**, *21*, 735. [CrossRef]

312. Vaquero, F.; Navarro, R.; Fierro, J. Evolution of the nanostructure of CdS using solvothermal synthesis at different temperature and its influence on the photoactivity for hydrogen production. *Int. J. Hydrogen Energy* **2016**, *41*, 11558–11567. [CrossRef]

313. Sun, S.; Gao, P.; Yang, Y.; Yang, P.; Chen, Y.; Wang, Y. N-doped TiO_2 nanobelts with coexposed (001) and (101) facets and their highly efficient visible-light-driven photocatalytic hydrogen production. *ACS Appl. Mater. Interfaces* **2016**, *8*, 18126–18131. [CrossRef]

314. Wang, F.; Jin, Z.; Jiang, Y.; Backus, E.H.; Bonn, M.; Lou, S.N.; Turchinovich, D.; Amal, R. Probing the charge separation process on In_2S_3/Pt-TiO_2 nanocomposites for boosted visible-light photocatalytic hydrogen production. *Appl. Catal. B Environ.* **2016**, *198*, 25–31. [CrossRef]

315. Wang, H.; Li, Y.; Shu, D.; Chen, X.; Liu, X.; Wang, X.; Zhang, J.; Wang, H. CoPtx-loaded $Zn_{0.5}Cd_{0.5}S$ nanocomposites for enhanced visible light photocatalytic H_2 production. *Int. J. Energy Res.* **2016**, *40*, 1280–1286. [CrossRef]

316. Wang, J.; Chen, Y.; Zhou, W.; Tian, G.; Xiao, Y.; Fu, H.; Fu, H. Cubic quantum dot/hexagonal microsphere $ZnIn_2S_4$ heterophase junctions for exceptional visible-light-driven photocatalytic H_2 evolution. *J. Mater. Chem. A* **2017**, *5*, 8451–8460. [CrossRef]

317. Wang, J.; Wang, Z.; Zhu, Z. Synergetic effect of $Ni(OH)_2$ cocatalyst and CNT for high hydrogen generation on CdS quantum dot sensitized TiO_2 photocatalyst. *Appl. Catal. B Environ.* **2017**, *204*, 577–583. [CrossRef]

318. Wang, L.; Di, Q.; Sun, M.; Liu, J.; Cao, C.; Liu, J.; Xu, M.; Zhang, J. Assembly-promoted photocatalysis: Three-dimensional assembly of CdS_xSe_{1-x} (x = 0–1) quantum dots into nanospheres with enhanced photocatalytic performance. *J. Mater.* **2017**, *3*, 63–70. [CrossRef]

319. Wang, Y.; Zhang, Y.; Jiang, Z.; Jiang, G.; Zhao, Z.; Wu, Q.; Liu, Y.; Xu, Q.; Duan, A.; Xu, C. Controlled fabrication and enhanced visible-light photocatalytic hydrogen production of Au@ CdS/MIL-101 heterostructure. *Appl. Catal. B Environ.* **2016**, *185*, 307–314. [CrossRef]

320. Wang, Z.; Wang, S.; Liu, J.; Jiang, W.; Zhou, Y.; An, C.; Zhang, J. Synthesis of $AgInS_2$-xAg_2S-$yZnS$-zIn_6S_7 (x, y, z = 0, or 1) Nanocomposites with Composition-Dependent Activity towards Solar Hydrogen Evolution. *Materials* **2016**, *9*, 329. [CrossRef]

321. Wu, A.; Tian, C.; Jiao, Y.; Yan, Q.; Yang, G.; Fu, H. Sequential two-step hydrothermal growth of MoS_2/CdS core-shell heterojunctions for efficient visible light-driven photocatalytic H_2 evolution. *Appl. Catal. B Environ.* **2017**, *203*, 955–963. [CrossRef]

322. Wu, L.; Gong, J.; Ge, L.; Han, C.; Fang, S.; Xin, Y.; Li, Y.; Lu, Y. AuPd bimetallic nanoparticles decorated $Cd_{0.5}Zn_{0.5}S$ photocatalysts with enhanced visible-light photocatalytic H_2 production activity. *Int. J. Hydrogen Energy* **2016**, *41*, 14704–14712. [CrossRef]

323. Xia, Y.; Li, Q.; Lv, K.; Tang, D.; Li, M. Superiority of graphene over carbon analogs for enhanced photocatalytic H_2-production activity of $ZnIn_2S_4$. *Appl. Catal. B Environ.* **2017**, *206*, 344–352. [CrossRef]

324. Xin, Y.; Lu, Y.; Han, C.; Ge, L.; Qiu, P.; Li, Y.; Fang, S. Novel NiS cocatalyst decorating ultrathin 2D TiO_2 nanosheets with enhanced photocatalytic hydrogen evolution activity. *Mater. Res. Bull.* **2017**, *87*, 123–129. [CrossRef]

325. Xing, Z.; Zong, X.; Zhu, Y.; Chen, Z.; Bai, Y.; Wang, L. A nanohybrid of CdTe@ CdS nanocrystals and titania nanosheets with p–n nanojunctions for improved visible light-driven hydrogen production. *Catal. Today* **2016**, *264*, 229–235. [CrossRef]

326. Yan, J.; Li, X.; Yang, S.; Wang, X.; Zhou, W.; Fang, Y.; Zhang, S.; Peng, F.; Zhang, S. Design and preparation of CdS/H-3D-TiO_2/Pt-wire photocatalysis system with enhanced visible-light driven H_2 evolution. *Int. J. Hydrogen Energy* **2017**, *42*, 928–937. [CrossRef]

327. Yan, Q.; Wu, A.; Yan, H.; Dong, Y.; Tian, C.; Jiang, B.; Fu, H. Gelatin-assisted synthesis of ZnS hollow nanospheres: The microstructure tuning, formation mechanism and application for Pt-free photocatalytic hydrogen production. *CrystEngComm* **2017**, *19*, 461–468. [CrossRef]

328. Yang, L.; Guo, S.; Li, X. Au nanoparticles@ MoS_2 core-shell structures with moderate MoS_2 coverage for efficient photocatalytic water splitting. *J. Alloys Compd.* **2017**, *706*, 82–88. [CrossRef]

329. Yang, Y.; Zhang, Y.; Fang, Z.; Zhang, L.; Zheng, Z.; Wang, Z.; Feng, W.; Weng, S.; Zhang, S.; Liu, P. Simultaneous Realization of Enhanced Photoactivity and Promoted Photostability by Multilayered MoS_2 Coating on CdS Nanowire Structure via Compact Coating Methodology. *ACS Appl. Mater. Interfaces* **2017**, *9*, 6950–6958. [CrossRef]

330. Yu, X.; Shi, J.; Wang, L.; Wang, W.; Bian, J.; Feng, L.; Li, C. A novel Au NPs-loaded MoS_2/RGO composite for efficient hydrogen evolution under visible light. *Mater. Lett.* **2016**, *182*, 125–128. [CrossRef]

331. Yuan, Y.J.; Chen, D.Q.; Huang, Y.W.; Yu, Z.T.; Zhong, J.S.; Chen, T.T.; Tu, W.G.; Guan, Z.J.; Cao, D.P.; Zou, Z.G. MoS$_2$ Nanosheet-Modified CuInS$_2$ Photocatalyst for Visible-Light-Driven Hydrogen Production from Water. *ChemSusChem* **2016**, *9*, 1003–1009. [CrossRef]

332. Yuan, Y.-J.; Tu, J.-R.; Ye, Z.-J.; Chen, D.-Q.; Hu, B.; Huang, Y.-W.; Chen, T.-T.; Cao, D.-P.; Yu, Z.-T.; Zou, Z.-G. MoS$_2$-graphene/ZnIn$_2$S$_4$ hierarchical microarchitectures with an electron transport bridge between light-harvesting semiconductor and cocatalyst: A highly efficient photocatalyst for solar hydrogen generation. *Appl. Catal. B Environ.* **2016**, *188*, 13–22. [CrossRef]

333. Yue, Z.; Liu, A.; Zhang, C.; Huang, J.; Zhu, M.; Du, Y.; Yang, P. Noble-metal-free hetero-structural CdS/Nb$_2$O$_5$/N-doped-graphene ternary photocatalytic system as visible-light-driven photocatalyst for hydrogen evolution. *Appl. Catal. B Environ.* **2017**, *201*, 202–210. [CrossRef]

334. Zhang, J.; Yao, W.; Huang, C.; Shi, P.; Xu, Q. High efficiency and stable tungsten phosphide cocatalysts for photocatalytic hydrogen production. *J. Mater. Chem. A* **2017**, *5*, 12513–12519. [CrossRef]

335. Zhang, N.; Chen, D.; Cai, B.; Wang, S.; Niu, F.; Qin, L.; Huang, Y. Facile synthesis of CdS ZnWO$_4$ composite photocatalysts for efficient visible light driven hydrogen evolution. *Int. J. Hydrogen Energy* **2017**, *42*, 1962–1969. [CrossRef]

336. Zhang, S.; Wang, L.; Zeng, Y.; Xu, Y.; Tang, Y.; Luo, S.; Liu, Y.; Liu, C. CdS-Nanoparticles-Decorated Perpendicular Hybrid of MoS$_2$ and N-Doped Graphene Nanosheets for Omnidirectional Enhancement of Photocatalytic Hydrogen Evolution. *ChemCatChem* **2016**, *8*, 2557–2564. [CrossRef]

337. Zhang, Y.; Han, L.; Wang, C.; Wang, W.; Ling, T.; Yang, J.; Dong, C.; Lin, F.; Du, X.-W. Zinc-Blende CdS Nanocubes with Coordinated Facets for Photocatalytic Water Splitting. *ACS Catal.* **2017**, *7*, 1470–1477. [CrossRef]

338. Zhao, H.; Sun, R.; Li, X.; Sun, X. Enhanced photocatalytic activity for hydrogen evolution from water by Zn$_{0.5}$Cd$_{0.5}$S/WS$_2$ heterostructure. *Mater. Sci. Semicond. Process.* **2017**, *59*, 68–75. [CrossRef]

339. Zhou, X.; Huang, J.; Zhang, H.; Sun, H.; Tu, W. Controlled synthesis of CdS nanoparticles and their surface loading with MoS 2 for hydrogen evolution under visible light. *Int. J. Hydrogen Energy* **2016**, *41*, 14758–14767. [CrossRef]

340. Jiang, D.; Chen, X.; Zhang, Z.; Zhang, L.; Wang, Y.; Sun, Z.; Irfan, R.M.; Du, P. Highly efficient simultaneous hydrogen evolution and benzaldehyde production using cadmium sulfide nanorods decorated with small cobalt nanoparticles under visible light. *J. Catal.* **2018**, *357*, 147–153. [CrossRef]

341. Kumar, D.P.; Park, H.; Kim, E.H.; Hong, S.; Gopannagari, M.; Reddy, D.A.; Kim, T.K. Noble metal-free metal-organic framework-derived onion slice-type hollow cobalt sulfide nanostructures: Enhanced activity of CdS for improving photocatalytic hydrogen production. *Appl. Catal. B Environ.* **2018**, *224*, 230–238. [CrossRef]

342. Lv, J.-X.; Zhang, Z.-M.; Wang, J.; Lu, X.-L.; Zhang, W.; Lu, T.-B. In situ synthesis of CdS/graphdiyne heterojunction for enhanced photocatalytic activity of hydrogen production. *ACS Appl. Mater. Interfaces* **2019**, *11*, 2655–2661. [CrossRef]

343. Feng, C.; Chen, Z.; Hou, J.; Li, J.; Li, X.; Xu, L.; Sun, M.; Zeng, R. Effectively enhanced photocatalytic hydrogen production performance of one-pot synthesized MoS$_2$ clusters/CdS nanorod heterojunction material under visible light. *Chem. Eng. J.* **2018**, *345*, 404–413. [CrossRef]

344. Liu, Y.; Ma, Y.; Liu, W.; Shang, Y.; Zhu, A.; Tan, P.; Xiong, X.; Pan, J. Facet and morphology dependent photocatalytic hydrogen evolution with CdS nanoflowers using a novel mixed solvothermal strategy. *J. Colloid Interface Sci.* **2018**, *513*, 222–230. [CrossRef]

345. Wang, L.; Xu, N.; Pan, X.; He, Y.; Wang, X.; Su, W. Cobalt lactate complex as a hole cocatalyst for significantly enhanced photocatalytic H$_2$ production activity over CdS nanorods. *Catal. Sci. Technol.* **2018**, *8*, 1599–1605. [CrossRef]

346. Abe, R. Recent progress on photocatalytic and photoelectrochemical water splitting under visible light irradiation. *J. Photochem. Photobiol. C Photochem. Rev.* **2010**, *11*, 179–209. [CrossRef]

347. Ahmed, A.Y.; Kandiel, T.A.; Ivanova, I.; Bahnemann, D. Photocatalytic and photoelectrochemical oxidation mechanisms of methanol on TiO$_2$ in aqueous solution. *Appl. Surf. Sci.* **2014**, *319*, 44–49. [CrossRef]

348. Al-Ahmed, A.; Mukhtar, B.; Hossain, S.; Javaid Zaidi, S.; Rahman, S. Application of Titanium dioxide (TiO$_2$) based photocatalytic nanomaterials in Solar and Hydrogen Energy: A Short Review. *Mater. Sci. Forum* **2012**, *712*, 25–47. [CrossRef]

349. Amao, Y. Solar fuel production based on the artificial photosynthesis system. *ChemCatChem* **2011**, *3*, 458–474. [CrossRef]

350. An, X.; Jimmy, C.Y. Graphene-based photocatalytic composites. *RSC Adv.* **2011**, *1*, 1426–1434. [CrossRef]

351. Ashokkumar, M. An overview on semiconductor particulate systems for photoproduction of hydrogen. *Int. J. Hydrogen Energy* **1998**, *23*, 427–438. [CrossRef]

352. Babu, V.J.; Vempati, S.; Uyar, T.; Ramakrishna, S. Review of one-dimensional and two-dimensional nanostructured materials for hydrogen generation. *Phys. Chem. Chem. Phys.* **2015**, *17*, 2960–2986. [CrossRef]

353. Bai, S.; Yin, W.; Wang, L.; Li, Z.; Xiong, Y. Surface and interface design in cocatalysts for photocatalytic water splitting and CO_2 reduction. *RSC Adv.* **2016**, *6*, 57446–57463. [CrossRef]

354. Bowker, M. Sustainable hydrogen production by the application of ambient temperature photocatalysis. *Green Chem.* **2011**, *13*, 2235–2246. [CrossRef]

355. Chen, X.; Li, C.; Grätzel, M.; Kostecki, R.; Mao, S.S. Nanomaterials for renewable energy production and storage. *Chem. Soc. Rev.* **2012**, *41*, 7909–7937. [CrossRef]

356. Colmenares, J.C.; Luque, R. Heterogeneous photocatalytic nanomaterials: Prospects and challenges in selective transformations of biomass-derived compounds. *Chem. Soc. Rev.* **2014**, *43*, 765–778. [CrossRef] [PubMed]

357. Colón, G. Towards the hydrogen production by photocatalysis. *Appl. Catal. A Gen.* **2016**, *518*, 48–59. [CrossRef]

358. Fang, W.; Xing, M.; Zhang, J. Modifications on reduced titanium dioxide photocatalysts: A review. *J. Photochem. Photobiol. C Photochem. Rev.* **2017**, *32*, 21–39. [CrossRef]

359. Fornasiero, P.; Christoforidis, K.C. Photocatalytic Hydrogen production: A rift into the future energy supply. *ChemCatChem* **2017**, *9*, 1523–1544.

360. Fresno, F.; Portela, R.; Suárez, S.; Coronado, J.M. Photocatalytic materials: Recent achievements and near future trends. *J. Mater. Chem. A* **2014**, *2*, 2863–2884. [CrossRef]

361. Gholipour, M.R.; Dinh, C.-T.; Béland, F.; Do, T.-O. Nanocomposite heterojunctions as sunlight-driven photocatalysts for hydrogen production from water splitting. *Nanoscale* **2015**, *7*, 8187–8208. [CrossRef]

362. Grabowska, E. Selected perovskite oxides: Characterization, preparation and photocatalytic properties—A review. *Appl. Catal. B Environ.* **2016**, *186*, 97–126. [CrossRef]

363. Guo, L.; Jing, D.; Liu, M.; Chen, Y.; Shen, S.; Shi, J.; Zhang, K. Functionalized nanostructures for enhanced photocatalytic performance under solar light. *Beilstein J. Nanotechnol.* **2014**, *5*, 994–1004. [CrossRef]

364. Han, B.; Hu, Y.H. MoS_2 as a co-catalyst for photocatalytic hydrogen production from water. *Energy Sci. Eng.* **2016**, *4*, 285–304. [CrossRef]

365. Hisatomi, T.; Kubota, J.; Domen, K. Recent advances in semiconductors for photocatalytic and photoelectrochemical water splitting. *Chem. Soc. Rev.* **2014**, *43*, 7520–7535. [CrossRef] [PubMed]

366. Jiao, W.; Shen, W.; Rahman, Z.U.; Wang, D. Recent progress in red semiconductor photocatalysts for solar energy conversion and utilization. *Nanotechnol. Rev.* **2016**, *5*, 135–145. [CrossRef]

367. Junge, H.; Rockstroh, N.; Fischer, S.; Brückner, A.; Ludwig, R.; Lochbrunner, S.; Kühn, O.; Beller, M. Light to Hydrogen: Photocatalytic Hydrogen Generation from Water with Molecularly-Defined Iron Complexes. *Inorganics* **2017**, *5*, 14. [CrossRef]

368. Kagkoura, A.; Skaltsas, T.; Tagmatarchis, N. Transition metal chalcogenides/graphene ensembles for light-induced energy applications. *Chem. Eur. J.* **2017**, *23*, 12967–12979. [CrossRef]

369. Kitano, M.; Tsujimaru, K.; Anpo, M. Hydrogen production using highly active titanium oxide-based photocatalysts. *Top. Catal.* **2008**, *49*, 4. [CrossRef]

370. Kudo, A.; Miseki, Y. Heterogeneous photocatalyst materials for water splitting. *Chem. Soc. Rev.* **2009**, *38*, 253–278. [CrossRef]

371. Kumar, P.S.; Sundaramurthy, J.; Sundarrajan, S.; Babu, V.J.; Singh, G.; Allakhverdiev, S.I.; Ramakrishna, S. Hierarchical electrospun nanofibers for energy harvesting, production and environmental remediation. *Energy Environ. Sci.* **2014**, *7*, 3192–3222. [CrossRef]

372. Leung, D.Y.; Fu, X.; Wang, C.; Ni, M.; Leung, M.K.; Wang, X.; Fu, X. Hydrogen Production over Titania-Based Photocatalysts. *ChemSusChem* **2010**, *3*, 681–694. [CrossRef] [PubMed]

373. Li, C.; Xu, Y.; Tu, W.; Chen, G.; Xu, R. Metal-free photocatalysts for various applications in energy conversion and environmental purification. *Green Chem.* **2017**, *19*, 882–899. [CrossRef]

374. Francesco, P.; Fabrizio, S.; Marco, M.; Claudio, M.; Valter, M. The Role of Surface Texture on the Photocatalytic H_2 Production on TiO_2. *Catalysts* **2019**, *9*, 32.

375. Zhang, X.; Wang, Y.; Liu, B.; Sang, Y.; Liu, H. Heterostructures construction on TiO_2 nanobelts: A powerful tool for building high-performance photocatalysts. *Appl. Catal. B Environ.* **2017**, *202*, 620–641. [CrossRef]

376. Li, X.; Yu, J.; Wageh, S.; Al-Ghamdi, A.A.; Xie, J. Graphene in photocatalysis: A review. *Small* **2016**, *12*, 6640–6696. [CrossRef] [PubMed]

377. Li, Y.; Li, Y.-L.; Sa, B.; Ahuja, R. Review of two-dimensional materials for photocatalytic water splitting from a theoretical perspective. *Catal. Sci. Technol.* **2017**, *7*, 545–559. [CrossRef]

378. Low, J.; Jiang, C.; Cheng, B.; Wageh, S.; Al-Ghamdi, A.A.; Yu, J. A Review of Direct Z-Scheme Photocatalysts. *Small Methods* **2017**, *1*, 1700080. [CrossRef]

379. Maeda, K. Photocatalytic water splitting using semiconductor particles: History and recent developments. *J. Photochem. Photobiol. C Photochem. Rev.* **2011**, *12*, 237–268. [CrossRef]

380. Martha, S.; Sahoo, P.C.; Parida, K. An overview on visible light responsive metal oxide based photocatalysts for hydrogen energy production. *RSC Adv.* **2015**, *5*, 61535–61553. [CrossRef]

381. Matsuoka, M.; Kitano, M.; Takeuchi, M.; Tsujimaru, K.; Anpo, M.; Thomas, J.M. Photocatalysis for new energy production: Recent advances in photocatalytic water splitting reactions for hydrogen production. *Catal. Today* **2007**, *122*, 51–61. [CrossRef]

382. Morales-Torres, S.; Pastrana-Martínez, L.M.; Figueiredo, J.L.; Faria, J.L.; Silva, A.M. Design of graphene-based TiO_2 photocatalysts—A review. *Environ. Sci. Pollut. Res.* **2012**, *19*, 3676–3687. [CrossRef] [PubMed]

383. Nguyen-Phan, T.-D.; Baber, A.E.; Rodriguez, J.A.; Senanayake, S.D. Au and Pt nanoparticle supported catalysts tailored for H_2 production: From models to powder catalysts. *Appl. Catal. A Gen.* **2016**, *518*, 18–47. [CrossRef]

384. Ni, M.; Leung, M.K.; Leung, D.Y.; Sumathy, K. A review and recent developments in photocatalytic water-splitting using TiO_2 for hydrogen production. *Renew. Sustain. Energy Rev.* **2007**, *11*, 401–425. [CrossRef]

385. Nurlaela, E.; Ziani, A.; Takanabe, K. Tantalum nitride for photocatalytic water splitting: Concept and applications. *Mater. Renew. Sustain. Energy* **2016**, *5*, 18. [CrossRef]

386. Pasternak, S.; Paz, Y. On the similarity and dissimilarity between photocatalytic water splitting and photocatalytic degradation of pollutants. *ChemPhysChem* **2013**, *14*, 2059–2070. [CrossRef]

387. Preethi, V.; Kanmani, S. Photocatalytic hydrogen production. *Mater. Sci. Semicond. Process.* **2013**, *16*, 561–575. [CrossRef]

388. Primo, A.; Corma, A.; García, H. Titania supported gold nanoparticles as photocatalyst. *Phys. Chem. Chem. Phys.* **2011**, *13*, 886–910. [CrossRef]

389. Protti, S.; Albini, A.; Serpone, N. Photocatalytic generation of solar fuels from the reduction of H_2O and CO_2: A look at the patent literature. *Phys. Chem. Chem. Phys.* **2014**, *16*, 19790–19827. [CrossRef]

390. Puga, A.V. Photocatalytic production of hydrogen from biomass-derived feedstocks. *Coord. Chem. Rev.* **2016**, *315*, 1–66. [CrossRef]

391. Ran, J.; Zhang, J.; Yu, J.; Jaroniec, M.; Qiao, S.Z. Earth-abundant cocatalysts for semiconductor-based photocatalytic water splitting. *Chem. Soc. Rev.* **2014**, *43*, 7787–7812. [CrossRef]

392. Samokhvalov, A. Hydrogen by photocatalysis with nitrogen codoped titanium dioxide. *Renew. Sustain. Energy Rev.* **2017**, *72*, 981–1000. [CrossRef]

393. Serrano, D.P.; Coronado, J.M.; Víctor, A.; Pizarro, P.; Botas, J.Á. Advances in the design of ordered mesoporous materials for low-carbon catalytic hydrogen production. *J. Mater. Chem. A* **2013**, *1*, 12016–12027. [CrossRef]

394. Sharma, P.; Kolhe, M.L. Review of sustainable solar hydrogen production using photon fuel on artificial leaf. *Int. J. Hydrogen Energy* **2017**, *42*, 22704–22712. [CrossRef]

395. Shi, J.; Guo, L. ABO_3-based photocatalysts for water splitting. *Prog. Nat. Sci. Mater. Int.* **2012**, *22*, 592–615. [CrossRef]

396. Shimura, K.; Yoshida, H. Heterogeneous photocatalytic hydrogen production from water and biomass derivatives. *Energy Environ. Sci.* **2011**, *4*, 2467–2481. [CrossRef]

397. Stroyuk, A.; Kryukov, A.; Kuchmii, S.Y.; Pokhodenko, V. Semiconductor photocatalytic systems for the production of hydrogen by the action of visible light. *Theor. Exp. Chem.* **2009**, *45*, 209. [CrossRef]

398. Wang, H.; Yuan, X.; Wu, Y.; Huang, H.; Peng, X.; Zeng, G.; Zhong, H.; Liang, J.; Ren, M. Graphene-based materials: Fabrication, characterization and application for the decontamination of wastewater and wastegas and hydrogen storage/generation. *Adv. Colloid Interface Sci.* **2013**, *195*, 19–40. [CrossRef]

399. Wang, H.; Zhang, L.; Chen, Z.; Hu, J.; Li, S.; Wang, Z.; Liu, J.; Wang, X. Semiconductor heterojunction photocatalysts: Design, construction, and photocatalytic performances. *Chem. Soc. Rev.* **2014**, *43*, 5234–5244. [CrossRef]

400. Wang, L. Strategies for efficient solar water splitting using carbon nitride. *Chem. Asian J.* **2017**, *12*, 1421–1434.

401. Wang, M.; Han, K.; Zhang, S.; Sun, L. Integration of organometallic complexes with semiconductors and other nanomaterials for photocatalytic H_2 production. *Coord. Chem. Rev.* **2015**, *287*, 1–14. [CrossRef]

402. Watanabe, M. Dye-sensitized photocatalyst for effective water splitting catalyst. *Sci. Technol. Adv. Mater.* **2017**, *18*, 705–723. [CrossRef]

403. Wen, M.; Mori, K.; Kuwahara, Y.; An, T.; Yamashita, H. Design and architecture of metal organic frameworks for visible light enhanced hydrogen production. *Appl. Catal. B Environ.* **2017**, *218*, 555–569. [CrossRef]

404. Xiao, F.X.; Miao, J.; Tao, H.B.; Hung, S.F.; Wang, H.Y.; Yang, H.B.; Chen, J.; Chen, R.; Liu, B. One-Dimensional Hybrid Nanostructures for Heterogeneous Photocatalysis and Photoelectrocatalysis. *Small* **2015**, *11*, 2115–2131. [CrossRef]

405. Xie, G.; Zhang, K.; Guo, B.; Liu, Q.; Fang, L.; Gong, J.R. Graphene-Based Materials for Hydrogen Generation from Light-Driven Water Splitting. *Adv. Mater.* **2013**, *25*, 3820–3839. [CrossRef]

406. Xing, J.; Fang, W.Q.; Zhao, H.J.; Yang, H.G. Inorganic photocatalysts for overall water splitting. *Chem. Asian J.* **2012**, *7*, 642–657. [CrossRef]

407. Xu, Y.; Xu, R. Nickel-based cocatalysts for photocatalytic hydrogen production. *Appl. Surf. Sci.* **2015**, *351*, 779–793. [CrossRef]

408. Xu, Y.; Zhang, B. Hydrogen photogeneration from water on the biomimetic hybrid artificial photocatalytic systems of semiconductors and earth-abundant metal complexes: Progress and challenges. *Catal. Sci. Technol.* **2015**, *5*, 3084–3096. [CrossRef]

409. Ye, S.; Wang, R.; Wu, M.-Z.; Yuan, Y.-P. A review on gC_3N_4 for photocatalytic water splitting and CO_2 reduction. *Appl. Surf. Sci.* **2015**, *358*, 15–27. [CrossRef]

410. Yin, S.; Han, J.; Zhou, T.; Xu, R. Recent progress in gC_3N_4 based low cost photocatalytic system: Activity enhancement and emerging applications. *Catal. Sci. Technol.* **2015**, *5*, 5048–5061. [CrossRef]

411. Yuan, Y.J.; Lu, H.W.; Yu, Z.T.; Zou, Z.G. Noble-Metal-Free Molybdenum Disulfide Cocatalyst for Photocatalytic Hydrogen Production. *ChemSusChem* **2015**, *8*, 4113–4127. [CrossRef]

412. Zhang, P.; Zhang, J.; Gong, J. Tantalum-based semiconductors for solar water splitting. *Chem. Soc. Rev.* **2014**, *43*, 4395–4422. [CrossRef]

413. Zhang, Q.; Gangadharan, D.T.; Liu, Y.; Xu, Z.; Chaker, M.; Ma, D. Recent advancements in plasmon-enhanced visible light-driven water splitting. *J. Mater.* **2017**, *3*, 33–50. [CrossRef]

414. Zhang, X.; Peng, T.; Song, S. Recent advances in dye-sensitized semiconductor systems for photocatalytic hydrogen production. *J. Mater. Chem. A* **2016**, *4*, 2365–2402. [CrossRef]

415. Zhao, X.; Wang, P.; Long, M. Electro- and Photocatalytic Hydrogen Production by Molecular Cobalt Complexes with Pentadentate Ligands. *Comments Inorg. Chem.* **2017**, *37*, 238–270. [CrossRef]

416. Zhao, Z.; Sun, Y.; Dong, F. Graphitic carbon nitride based nanocomposites: A review. *Nanoscale* **2015**, *7*, 15–37. [CrossRef] [PubMed]

417. Zou, X.; Zhang, Y. Noble metal-free hydrogen evolution catalysts for water splitting. *Chem. Soc. Rev.* **2015**, *44*, 5148–5180. [CrossRef] [PubMed]

418. Shwetharani, R.; Sakar, M.; Fernando, C.; Binas, V.; Balakrishna, R.G. Recent advances and strategies to tailor the energy levels, active sites and electron mobility in titania and its doped/composite analogues for hydrogen evolution in sunlight. *Catal. Sci. Technol.* **2019**, *9*, 12–46. [CrossRef]

419. Yuan, Y.-J.; Chen, D.; Yu, Z.-T.; Zou, Z.-G. Cadmium sulfide-based nanomaterials for photocatalytic hydrogen production. *J. Mater. Chem. A* **2018**, *6*, 11606–11630. [CrossRef]

420. Yuan, Y.-J.; Ye, Z.-J.; Lu, H.-W.; Hu, B.; Li, Y.-H.; Chen, D.-Q.; Zhong, J.-S.; Yu, Z.-T.; Zou, Z.-G. Constructing anatase TiO_2 nanosheets with exposed (001) facets/layered MoS_2 two-dimensional nanojunctions for enhanced solar hydrogen generation. *ACS Catal.* **2015**, *6*, 532–541. [CrossRef]

421. Liu, S.-H.; Tang, W.-T.; Lin, W.-X. Self-assembled ionic liquid synthesis of nitrogen-doped mesoporous TiO_2 for visible-light-responsive hydrogen production. *Int. J. Hydrogen Energy* **2017**, *42*, 24006–24013. [CrossRef]

422. Chen, Z.; Jiang, X.; Zhu, C.; Shi, C. Chromium-modified $Bi_4Ti_3O_{12}$ photocatalyst: Application for hydrogen evolution and pollutant degradation. *Appl. Catal. B Environ.* **2016**, *199*, 241–251. [CrossRef]

423. Yang, S.; Wang, H.; Yu, H.; Zhang, S.; Fang, Y.; Zhang, S.; Peng, F. A facile fabrication of hierarchical Ag nanoparticles-decorated N-TiO$_2$ with enhanced photocatalytic hydrogen production under solar light. *Int. J. Hydrogen Energy* **2016**, *41*, 3446–3455. [CrossRef]

424. Silva, L.A.; Ryu, S.Y.; Choi, J.; Choi, W.; Hoffmann, M.R. Photocatalytic hydrogen production with visible light over Pt-interlinked hybrid composites of cubic-phase and hexagonal-phase CdS. *J. Phys. Chem. C* **2008**, *112*, 12069–12073. [CrossRef]

425. Qin, N.; Xiong, J.; Liang, R.; Liu, Y.; Zhang, S.; Li, Y.; Li, Z.; Wu, L. Highly efficient photocatalytic H$_2$ evolution over MoS$_2$/CdS-TiO$_2$ nanofibers prepared by an electrospinning mediated photodeposition method. *Appl. Catal. B Environ.* **2017**, *202*, 374–380. [CrossRef]

426. Gopannagari, M.; Kumar, D.P.; Reddy, D.A.; Hong, S.; Song, M.I.; Kim, T.K. In situ preparation of few-layered WS$_2$ nanosheets and exfoliation into bilayers on CdS nanorods for ultrafast charge carrier migrations toward enhanced photocatalytic hydrogen production. *J. Catal.* **2017**, *351*, 153–160. [CrossRef]

427. Wang, M.; Na, Y.; Gorlov, M.; Sun, L. Light-driven hydrogen production catalysed by transition metal complexes in homogeneous systems. *Dalton Trans.* **2009**, 6458–6467. [CrossRef] [PubMed]

428. Linkous, C.A.; Huang, C.; Fowler, J.R. UV photochemical oxidation of aqueous sodium sulfide to produce hydrogen and sulfur. *J. Photochem. Photobiol. A Chem.* **2004**, *168*, 153–160. [CrossRef]

429. Huang, C.; Linkous, C.A.; Adebiyi, O.; T-Raissi, A. Hydrogen production via photolytic oxidation of aqueous sodium sulfite solutions. *Environ. Sci. Technol.* **2010**, *44*, 5283–5288. [CrossRef]

430. Li, C.; Hu, P.; Meng, H.; Jiang, Z. Role of Sulfites in the Water Splitting Reaction. *J. Solut. Chem.* **2016**, *45*, 67–80. [CrossRef]

431. Husin, H.; Adisalamun, S.Y.; Asnawi, T.M.; Hasfita, F. Pt nanoparticle on La$_{0.02}$Na$_{0.98}$TaO$_3$ catalyst for hydrogen evolution from glycerol aqueous solution. *AIP Conf. Proc.* **2017**, *1788*, 030073.

432. López-Tenllado, F.; Hidalgo-Carrillo, J.; Montes, V.; Marinas, A.; Urbano, F.; Marinas, J.; Ilieva, L.; Tabakova, T.; Reid, F. A comparative study of hydrogen photocatalytic production from glycerol and propan-2-ol on M/TiO$_2$ systems (M = Au, Pt, Pd). *Catal. Today* **2017**, *280*, 58–64. [CrossRef]

433. Li, F.; Gu, Q.; Niu, Y.; Wang, R.; Tong, Y.; Zhu, S.; Zhang, H.; Zhang, Z.; Wang, X. Hydrogen evolution from aqueous-phase photocatalytic reforming of ethylene glycol over Pt/TiO$_2$ catalysts: Role of Pt and product distribution. *Appl. Surf. Sci.* **2017**, *391*, 251–258. [CrossRef]

434. Oscar, Q.C.; Socorro, O.R.; Solís-Gómezb, A.; Rosendo, L.; Ricardo, G. Enhanced photocatalytic hydrogen production by CdS nanofibers modified with graphene oxide and nickel nanoparticles under visible light. *Fuel* **2019**, *237*, 227–235.

435. Andrea, S.; Francesca, G.; Federica, M.; Michela, S.; Daniele, D.; Lorenzo, M.; Antonella, P. Photocatalytic hydrogen evolution assisted by aqueous (waste)biomass under simulated solar light: Oxidized g-C$_3$N$_4$ vs. P25 titanium dioxide. *Int. J. Hydrogen Energy* **2019**, *44*, 4072–4078.

436. Tao, C.; Jie, M.; Qingyun, L.; Xiao, W.; Jixue, L.; Ze, Z. One-step synthesis of hollow BaZrO$_3$ nanocrystals with oxygen vacancies for photocatalytic hydrogen evolution from pure water. *J. Alloys Compd.* **2019**, *780*, 498–503.

437. Bahruji, H.; Bowker, M.; Davies, P.R.; Pedrono, F. New insights into the mechanism of photocatalytic reforming on Pd/TiO$_2$. *Appl. Catal. B Environ.* **2011**, *107*, 205–209. [CrossRef]

438. Fu, X.; Wang, X.; Leung, D.Y.; Gu, Q.; Chen, S.; Huang, H. Photocatalytic reforming of C$_3$-polyols for H$_2$ production: Part (I). Role of their OH groups. *Appl. Catal. B Environ.* **2011**, *106*, 681–688. [CrossRef]

439. Shkrob, I.A.; Sauer, M.C.; Gosztola, D. Efficient, rapid photooxidation of chemisorbed polyhydroxyl alcohols and carbohydrates by TiO$_2$ nanoparticles in an aqueous solution. *J. Phys. Chem. B* **2004**, *108*, 12512–12517. [CrossRef]

440. Shkrob, I.A.; Sauer, M.C. Hole Scavenging and Photo-Stimulated Recombination of Electron—Hole Pairs in Aqueous TiO$_2$ Nanoparticles. *J. Phys. Chem. B* **2004**, *108*, 12497–12511. [CrossRef]

441. Du, M.-H.; Feng, J.; Zhang, S. Photo-oxidation of polyhydroxyl molecules on TiO$_2$ surfaces: From hole scavenging to light-induced self-assembly of TiO$_2$-cyclodextrin wires. *Phys. Rev. Lett.* **2007**, *98*, 066102. [CrossRef]

442. Wang, B.; Zhang, J.; Huang, F. Enhanced visible light photocatalytic H$_2$ evolution of metal-free g-C$_3$N$_4$/SiC heterostructured photocatalysts. *Appl. Surf. Sci.* **2017**, *391*, 449–456. [CrossRef]

443. Zhang, Z.; Zhang, Y.; Lu, L.; Si, Y.; Zhang, S.; Chen, Y.; Dai, K.; Duan, P.; Duan, L.; Liu, J. Graphitic carbon nitride nanosheet for photocatalytic hydrogen production: The impact of morphology and element composition. *Appl. Surf. Sci.* **2017**, *391*, 369–375. [CrossRef]

444. Fang, L.J.; Wang, X.L.; Li, Y.H.; Liu, P.F.; Wang, Y.L.; Zeng, H.D.; Yang, H.G. Nickel nanoparticles coated with graphene layers as efficient co-catalyst for photocatalytic hydrogen evolution. *Appl. Catal. B Environ.* **2017**, *200*, 578–584. [CrossRef]

445. Yuan, Y.J.; Shen, Z.; Wu, S.; Su, Y.; Pei, L.; Ji, Z.; Ding, M.; Bai, W.; Chen, Y.; Yu, Z.T.; et al. Liquid exfoliation of g-C$_3$N$_4$ nanosheets to construct 2D-2D MoS$_2$/g-C$_3$N$_4$ photocatalyst for enhanced photocatalytic H$_2$ production activity. *Appl. Catal. B Environ.* **2019**, *246*, 120–128. [CrossRef]

446. Cai, J.; Shen, J.; Zhang, X.; Ng, Y.H.; Huang, J.; Guo, W.; Lin, C.; Lai, Y. Light-Driven Sustainable Hydrogen Production Utilizing TiO$_2$ Nanostructures: A Review. *Small* **2019**, *3*, 1800184. [CrossRef]

447. Wang, C.; Wang, L.; Jin, J.; Liu, J.; Li, Y.; Wu, M.; Chen, L.; Wang, B.; Yang, X.; Su, B.-L. Probing effective photocorrosion inhibition and highly improved photocatalytic hydrogen production on monodisperse PANI@ CdS core-shell nanospheres. *Appl. Catal. B Environ.* **2016**, *188*, 351–359. [CrossRef]

448. Song, J.; Zhao, H.; Sun, R.; Li, X.; Sun, D. An efficient hydrogen evolution catalyst composed of palladium phosphorous sulphide (PdP ~ 0.33 S ~ 1.67) and twin nanocrystal Zn$_{0.5}$Cd$_{0.5}$S solid solution with both homo-and hetero-junctions. *Energy Environ. Sci.* **2017**, *10*, 225–235. [CrossRef]

449. Ma, S.; Xie, J.; Wen, J.; He, K.; Li, X.; Liu, W.; Zhang, X. Constructing 2D layered hybrid CdS nanosheets/MoS$_2$ heterojunctions for enhanced visible-light photocatalytic H$_2$ generation. *Appl. Surf. Sci.* **2017**, *391*, 580–591. [CrossRef]

450. Cheng, F.; Yin, H.; Xiang, Q. Low-temperature solid-state preparation of ternary CdS/g-C$_3$N$_4$/CuS nanocomposites for enhanced visible-light photocatalytic H$_2$-production activity. *Appl. Surf. Sci.* **2017**, *391*, 432–439. [CrossRef]

451. Tian, F.; Hou, D.; Hu, F.; Xie, K.; Qiao, X.; Li, D. Pouous TiO$_2$ nanofibers decorated CdS nanoparticles by SILAR method for enhanced visible-light-driven photocatalytic activity. *Appl. Surf. Sci.* **2017**, *391*, 295–302. [CrossRef]

452. Vignesh, K.; Suganthi, A.; Min, B.-K.; Kang, M. Photocatalytic activity of magnetically recoverable MnFe$_2$O$_4$/g-C$_3$N$_4$/TiO$_2$ nanocomposite under simulated solar light irradiation. *J. Mol. Catal. A Chem.* **2014**, *395*, 373–383. [CrossRef]

453. Zhen, W.; Ning, X.; Yang, B.; Wu, Y.; Li, Z.; Lu, G. The enhancement of CdS photocatalytic activity for water splitting via anti-photocorrosion by coating Ni$_2$P shell and removing nascent formed oxygen with artificial gill. *Appl. Catal. B Environ.* **2018**, *221*, 243–257. [CrossRef]

 *catalysts*

Article

Enhancement of Hydrogen Productions by Accelerating Electron-Transfers of Sulfur Defects in the CuS@CuGaS$_2$ Heterojunction Photocatalysts

Namgyu Son, Jun Neoung Heo, Young-Sang Youn, Youngsoo Kim, Jeong Yeon Do * and Misook Kang *

Department of Chemistry, College of Science, Yeungnam University, Gyeongsan, Gyeongbuk 38541, Korea; sng1107@naver.com (N.S.); hjn2521@naver.com (J.N.H.); ysyoun@yu.ac.kr (Y.-S.Y.); kimys6553@yu.ac.kr (Y.K.)
* Correspondence: daengi77@ynu.ac.kr (J.Y.D.); mskang@ynu.ac.kr (M.K.); Tel.: +82-53-810-3798 (J.Y.D.); +82-53-810-2363 (M.K.); Fax: +82-53-815-5412 (M.K.)

Received: 29 November 2018; Accepted: 28 December 2018; Published: 4 January 2019

Abstract: CuS and CuGaS$_2$ heterojunction catalysts were used to improve hydrogen production performance by photo splitting of methanol aqueous solution in the visible region in this study. CuGaS$_2$, which is a chalcogenide structure, can form structural defects to promote separation of electrons and holes and improve visible light absorbing ability. The optimum catalytic activity of CuGaS$_2$ was investigated by varying the heterojunction ratio of CuGaS$_2$ with CuS. Physicochemical properties of CuS, CuGaS$_2$ and CuS@CuGaS$_2$ nanoparticles were confirmed by X-ray diffraction, ultraviolet visible spectroscopy, high-resolution transmission electron microscopy, scanning electron microscopy and energy dispersive X-ray spectroscopy. Compared with pure CuS, the hydrogen production performance of CuGaS$_2$ doped with Ga dopant was improved by methanol photolysis, and the photoactivity of the heterogeneous CuS@CuGaS$_2$ catalyst was increased remarkably. Moreover, the 0.5CuS@1.5CuGaS$_2$ catalyst produced 3250 µmol of hydrogen through photolysis of aqueous methanol solution under 10 h UV light irradiation. According to the intensity modulated photovoltage spectroscopy (IMVS) results, the high photoactivity of the CuS@CuGaS$_2$ catalyst is attributed to the inhibition of recombination between electron-hole pairs, accelerating electron-transfer by acting as a trap site at the interface between CuGaS$_2$ structural defects and the heterojunction.

Keywords: hydrogen production; methanol photo-splitting; heterojunction; CuS@CuGaS$_2$; electron-hole recombination

---

1. Introduction

Copper sulfide (Cu$_{2-x}$S, 0< x < 1), a non-toxic and conductive chalcogen compound, has been continuously noted for its excellent photoelectric behavior, potential thermal/electrical properties, and unique biomedical properties for decades, and much extensive research on Cu$_{2-x}$S micro/nano structures is still being actively conducted. In particular, micro/nanostructured Cu$_{2-x}$S with well-controlled shapes, sizes, structures and compositions have already been applied as photocatalytic materials [1], energy conversion materials [2], biosensing materials [3], and bioimaging materials [4] and have shown reasonable results. However, comprehensive reviews of the Cu$_{2-x}$S structure in-depth in applications are still lacking. Therefore, it is necessary to categorize new functions or orientations of Cu$_{2-x}$S-based nanocomposites and to develop and improve their essential elements for specific applications. Many researchers have already published a number of strategies for synthesizing 0D (dimension) , 1D, 2D, and 3D micro/nanostructures (including polyhedra) [5–7], and their efforts have made important progress in identifying Cu$_{2-x}$S micro/nanostructures. Furthermore, improved Cu$_{2-x}$S composites with hollow structures or super-lattices could be extended to a variety of applications in terms of performance [8]. $Cu_{2-x}S$ belongs to the covellite mineral group with a hexagonal crystal structure, and belongs to the crystal group of P63/mmc. However, bond lengths and angles can be varied in various ways depending on the oxidation state of the copper or other anion exchanges instead of S^{2-} present in the surroundings. For example, $Cu_{2-x}S$ is very different from $Cu_{2-x}O$ but shows a similar structure to $Cu_{2-x}Se$ (klockmannite) [9]. CuS compound is paramagnetic due to the $3d^9$ electron arrangement, and some studies have reported that all Cu atoms in CuS have an oxidation state of Cu^+ based on XPS results [10]. However, it can be attributed to $(Cu^+)_2Cu^{2+}S_2$ with both Cu(I) and Cu(II) in the XRD crystal structure results [11]. There are many applications of $Cu_{2-x}S$ as a photocatalyst, among which Saranya et al. suggested that the morphology of CuS was influenced by the reaction time and surfactant, and its photocatalytic activity for decolorization of methylene blue (MB) dye under visible-light irradiation was 87% [12]. However, CuS is mainly used in combination with other types of photocatalysts rather than alone [13,14]. On the other hand, substitution of $Cu_{2-x}S$ with other metal ions instead of Cu(II) can produce a complex crystal structure with varying performance. For example, $CuInS_2$ and $CuGaS_2$, which are called CIS or CIGS, are used as light absorbers in thin film solar cells; they absorb visible light in a wide area and are stable to light, unlike $Cu_{2-x}S$. In particular, Salak et al. reported that $CuM^{3+}S_2$, a chalcogen compound, formed Cu defects on tetrahedrons and facilitated the separation of electrons and holes, which could maintain photoactivity for a long time [15]. Yue et al. concluded that $CuInS_2$ was most sensitive at 500 nm with an optimal apparent quantum yield of 23.85% [16]. Han et al. found that with an increase in the reaction temperature, the excitonic absorption peaks and band gap emission peaks were systematically red-shifted, thus exhibiting a quantum confinement effect, and the $CuGaS_2$ quantum dots showed promising visible-light-driven photocatalytic activity during degradation of rhodamine 6G [17]. We have already confirmed in previous studies that the $CuS@CuInS_2:In_2S_3$ catalyst has the ability to decompose water to produce hydrogen with high efficiency. Especially, it was found that the $CuInS_2$ layer inserted between CuS and In_2S_3 acts as an electron-rich interface to accelerate the reduction of water in this layer [18]. Furthermore, we have also found in previous studies that Ga has excellent performance in decomposing methanol to produce hydrogen, and that the hydrogen reverse-spillover phenomenon of Ga has a great influence on the removal of hydrogen from methanol [19]. Despite many previous studies, the excellent photosensitivity of chalcogen compounds, including $Cu_{2-x}S$, in multifunctional complexes is broadly applicable to a wide range of photochemical reactions, so research on chalcogen compounds remains of interest. In particular, if chalcogen compounds are expected to perform well in the photoreaction for hydrogen production from water decomposition and their photostability is guaranteed for a long time, development of the chalcogen photocatalyst will be quite a desirable area of study for the next generation of environmentally friendly energy sources.

Therefore, in this study, we applied two particles of CuS and $CuGaS_2$ as base catalysts and applied them to hydrogen production from water degradation by hetero-connecting them between two particles. The effect of the pure structure or heterojunction structure of the two particles on photoactivity was investigated. Five types of catalysts were prepared: CuS, $CuGaS_2$, $1.0CuS@1.0CuGaS_2$, $0.5CuS@1.5CuGaS_2$, and $1.5CuS@0.5CuGaS_2$. The molar ratios of the two particles at the heterojunction were $CuS:CuGaS_2$ = 1:1, 0.5:1.5, and 1.5:0.5, respectively, to determine which particles most influence catalytic activity.

2. Results and Discussion

Characteristics of CuS, $CuGaS_2$ and $CuS@CuGaS_2$ Nanoparticles

The XRD patterns (A) and high-resolution TEM (HRTEM) images (B) are shown in Figure 1 to confirm the crystallinity of synthesized CuS, $CuGaS_2$ and heterojunction $CuS@CuGaS_2$ nanoparticles. The main XRD peaks of CuS were observed at 2θ = 27.65° (101), 29.25° (102), 31.75° (103), 32.77° (006), 38.77° (105), 47.89° (110), 52.61° (108), 59.24 (116), 73.87° (208) and 78.96° (213) and were assigned to the covellite CuS of the hexagonal crystal structure (P63 / mmc space group, JCPDS card No. 01-078-0876) [20]. On the other hand, the peaks at 31.74° and 46.11° correspond to Cu_2S (JCPDS card no. 00-053-0522) of cubic crystal structure, meaning that the two structures are finely mixed [21]. The XRD patterns of the $CuGaS_2$ nanoparticle showed a $CuGaS_2$ peak with a tetragonal crystal structure (I-42d space group, JCPDS card no. 01-085-1574) [22], although CuS was mixed. The main XRD peaks of $CuGaS_2$ were assigned to 2θ = 29.11° (112), 33.49° (200), 48.63° (204) and 57.20° (312). Meanwhile, the XRD pattern of the heterojunction $CuS@CuGaS_2$ nanoparticles was very similar to the XRD pattern of the $CuGaS_2$ corresponding to the shell, but there was a slight difference in the intensity and position of the peaks as the ratio of $CuS:CuGaS_2$ varied. In particular, in the $0.5CuS@1.5CuGaS_2$ sample, the peak corresponding to the (112) crystal plane migrated at a higher angle than $CuGaS_2$. This is probably due to the small ionic radius of Ga^{3+} ions compared to Cu^{2+} [23], and it is expected that lattice parameters will be reduced according to Bragg's law [24], $n\lambda$ = 2d sin (θ) (where n is the order of reflection; the wavelength of the X-rays, d = the distance between two layers of the crystals, and θ = the angle of the incident light). It can be expected that lattice defects are formed as compared with pure CuS or $CuGaS_2$, since Ga^{2+} ions can be incorporated into the CuS lattice or the lattice gap in the process of heterojunction. This can act as an active site of the catalyst and enhance catalyst performance. Figure 1B) shows the high-resolution TEM (HRTEM) (a), the selected area electron diffraction (SAED) (b), and the elemental mapping image (c) of the heterojunction $0.5CuS@1.5CuGaS_2$ particles. This result not only shows the overall shape of the particles, but also explains the intrinsic crystal structure of the particles based on the lattice parameter values in relation to the XRD results. The $0.5CuS@1.5CuGaS_2$ particles are shown in polycrystalline form as aggregates of single crystals with different orientations, which are evidenced by the lattice images and the SAED patterns. In general, when a certain point in a SAED pattern is clearly marked, it signifies a single crystal, and when a continuous ring is drawn, it signifies a polycrystalline. Therefore, heterojunction $0.5CuS@1.5CuGaS_2$ particles appeared to be polycrystalline, and lattice patterns of CuS and $CuGaS_2$ were observed. We can expect CuS and $CuGaS_2$ to be bonded as shown by the difference in shading in the TEM image. A lattice corresponding to 0.288 nm (103 diffraction plane) and 0.305 nm (102 diffraction plane) of CuS was observed in the bright portion, and a lattice pattern of 0.266 nm (200 diffraction plane) and 0.188 nm (204 diffraction plane) of $CuGaS_2$ was observed in dark areas. In particular, the 200 diffraction plane of $CuGaS_2$ has a very distinct lattice pattern, and a clear diffraction spot was also observed in the SAED pattern. This is consistent with the XRD results of the heterojunction $0.5CuS@1.5CuGaS_2$ particles and demonstrates that the particle is a continuous polycrystalline structure with a ring pattern with partially defined diffraction spots. On the other hand, the element mapping (c) results show that the Cu, Ga and S elements are uniformly distributed in the $0.5CuS@1.5CuGaS_2$ particles, thus demonstrating the microstructure and composition of the heterogeneous bonded particles.

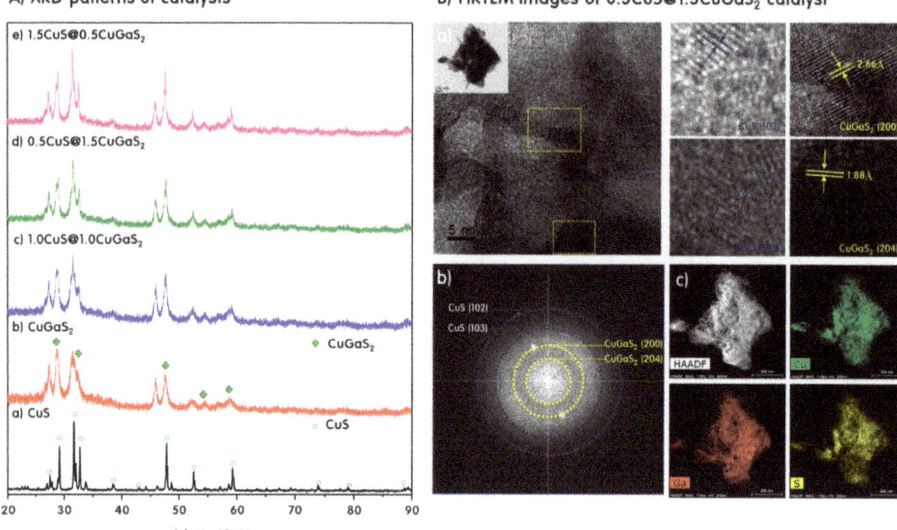

Figure 1. X-ray diffraction (XRD) patterns (**A**) and high-resolution transmission electron microscopy (HRTEM) images (**B**) of prepared samples.

Figure 2 shows the scanning electron microscope (SEM) image and energy-dispersive X-ray spectroscopy (EDS) analysis of synthesized CuS, $CuGaS_2$, and heterojunction $CuS@CuGaS_2$ nanoparticles, showing the composition of the components present on the surface of the particles. Figure 2A shows a SEM image of each sample, showing significant aggregation between the particles in the $CuGaS_2$ sample compared to CuS. On the other hand, the heterojunction $CuS@CuGaS_2$ sample inhibited the agglomeration of particles and the particle size became smaller. Figure 2B shows the EDS spectra of each sample and the atomic composition ratios are shown in the Table 1. Determination of the atomic ratio of the main metal species in the catalyst is very important because it relates to the density of the crystal lattice defects [25]. CuS and $CuGaS_2$, heterojunction $CuS@CuGaS_2$ particles show that Cu, Ga and S atomic components are precisely contained, and no other components are included. The atomic composition of pure CuS is close to the ideal stoichiometric mole fraction with a Cu:S ratio of 46.74:53.26. The composition of the $CuGaS_2$ sample was 31.46:27.26:41.27 with a Cu:Ga:S ratio slightly different from the stoichiometric ratio, but this can be predicted as a limitation of the EDS surface methodology. In addition, the heterojunction $CuS@CuGaS_2$ samples had a relatively low proportion of Ga. It is also expected that the difference in the sizes of Cu and Ga ions causes Ga to enter into the lattice of the crystal structure to reduce the amount of Ga exposed on the surface. Taking these factors into account, the overall molar ratio of Cu to Ga, S was almost quantitatively and reliably obtained.

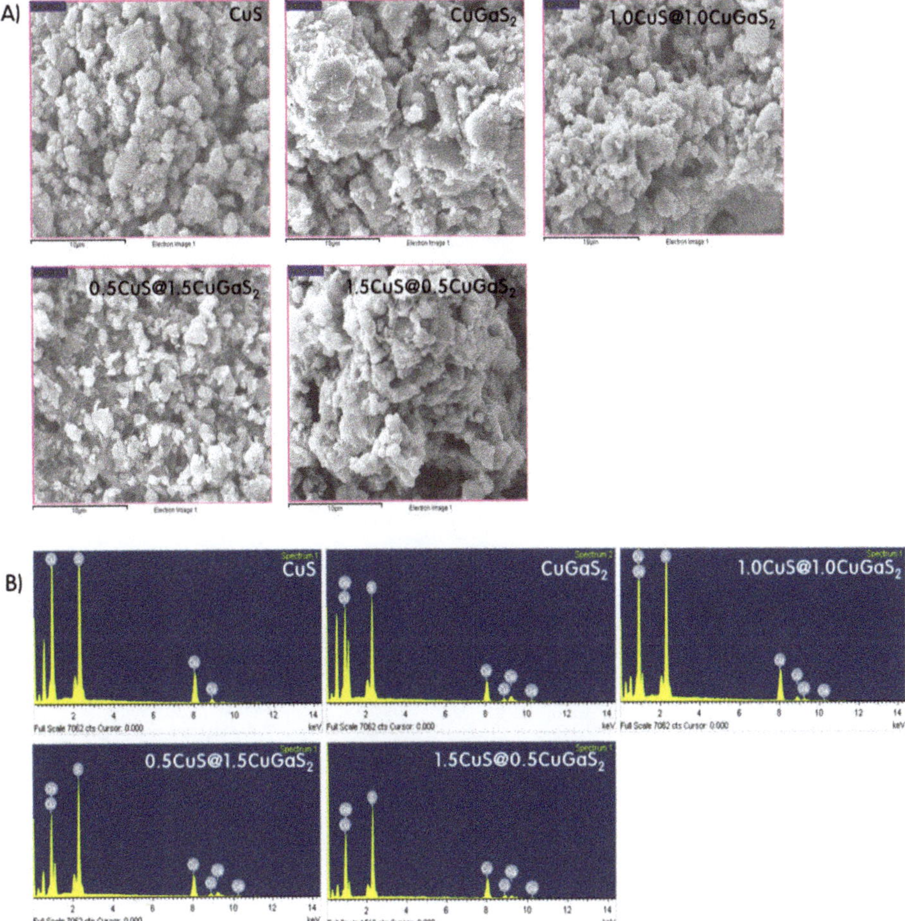

Figure 2. The SEM images (**A**) and energy-dispersive X-ray spectroscopy curves (**B**) of the CuS, CuGaS$_2$ and CuS@CuGaS$_2$ catalysts.

Table 1. Atomic compositions of the CuS, CuGaS$_2$ and CuS@CuGaS$_2$ catalysts determined by EDS.

Catalysts/Elements	Cu	Ga	S
CuS	46.74	–	53.26
CuGaS$_2$	31.46	27.26	41.27
1.0CuS@1.0CuGaS$_2$	32.99	23.06	43.95
0.5CuS@1.5CuGaS$_2$	15.00	36.58	48.43
1.5CuS@0.5CuGaS$_2$	41.05	18.31	40.64

Figure 3 shows the UV-visible reflectance spectra of the synthesized CuS, CuGaS$_2$ and heterojunction CuS@CuGaS$_2$ nanoparticles. UV-visible absorption spectroscopy is widely used to investigate microscopic changes caused by the chemical characteristics of the surface of the particles [26], and the optical band gap can be calculated through absorption spectra. Figure 3A clearly shows that the CuS, CuGaS$_2$ and heterojunction CuS@CuGaS$_2$ samples show absorption spectra in the region of 300~800 nm, and CuS in particular exhibited a pronounced absorption shoulder at about 620 nm. In the CuGaS$_2$ and CuS@CuGaS$_2$ samples, the apparent absorption shoulder peak disappeared, but the overall absorption range was similar to CuS. These results suggest that CuS, $CuGaS_2$ and $CuS@CuGaS_2$ samples can be used as promising photocatalytic materials to absorb visible light. Based on the absorption spectrum, the band gap was calculated by the Tauc equation [27], $\alpha h\nu = A (h\nu = E_g)^n$. Where $h\nu$ is the photon energy, α is the absorption coefficient, A is the constant relative material, and n is the value that depends on the transitional nature (2 is direct allowed transition, 2/3 is direct suppressed transition, 2/3 is indirectly permissible transition). In addition, the energy band gap can be predicted from the wavelength extrapolated from the exciton peak called $\lambda_{1/2}$, or the point where the end of the absorption curve meets the x axis. The band gaps of pure CuS samples and $CuGaS_2$ were 1.63 and 2.34 eV, respectively. This value is similar to the band gap reported in other papers [28]. The bandgap of the heterojunction $0.5CuS@1.5CuGaS_2$ particles was 2.32 eV. As $CuGaS_2$, which has a longer band gap, is bonded to CuS, $CuGaS_2$ first acts as a photosensitizer, and excited electrons can be transferred to CuS. If the bandgap is long, the recombination between the photogenerated electrons and the hole pair is delayed, and the photoactivity can be increased.

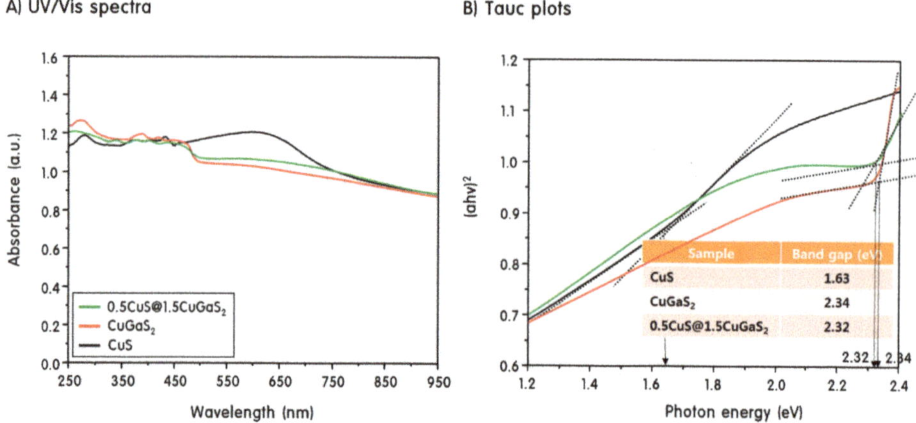

Figure 3. UV–visible spectroscopy curves (**A**) and Tauc plots (**B**) of CuS, $CuGaS_2$ and $CuS@CuGaS_2$ catalysts.

In general, the behavior of photogenerated carriers is closely related to photocatalytic activity. The generation, separation, transport and recombination of photogenerated electron-hole pairs have a great influence on photocatalytic activity [29]. Therefore, the photocurrent measurement and the photoluminescence measurement results are shown in Figure 4 in order to understand the separation efficiency and recombination characteristics of the photogenerated electrons and hole pairs. Figure 4A shows the results of measuring the photocurrent value when the light was irradiated by controlling the switch of the light source at intervals of 30 s. Electrons were excited by the irradiated light to generate excited electron and hole pairs. Their separation efficiency and mobility are closely related to the photocurrent value [30]. The photocurrent density values were increased in the order of CuS $< 1.5CuS@0.5CuGaS_2 < CuGaS_2 < 1.0CuS@1.0CuGaS_2 < 0.5CuS@1.5CuGaS_2$. The pure CuS catalyst showed only a slight tendency for photocurrent density to drop momentarily when the light was turned off. In general photocurrent results, the photogenerated holes migrate to the catalyst surface and are captured or trapped by the reduced species in the electrolyte, and the electrons undergo backside contact through the catalyst, leading to an increase in the initial anodic photocurrent. Then, after the competitive separation of the electron and hole pairs and the equilibrium of the recombination, the photocurrent is kept constant, while the CuS temporarily decreases the photocurrent. This is presumably due to the fact that the traced holes on the catalyst surface are not captured or trapped by the reduced species in the electrolyte, but instead are competitively recombined with the electrons in the conduction band of the catalyst [31]. On the other hand, the photocurrent density of $CuGaS_2$ and heterojunction $CuS@CuGaS_2$ catalysts was not only increased, but also showed excellent stability and reliability. It is believed that the addition of the Ga dopant induces more exciton formation and that structural or surface defects caused by heterojunction act as capture sites, or accelerate electron-transfer, resulting in more efficient photogenerated charge separation.

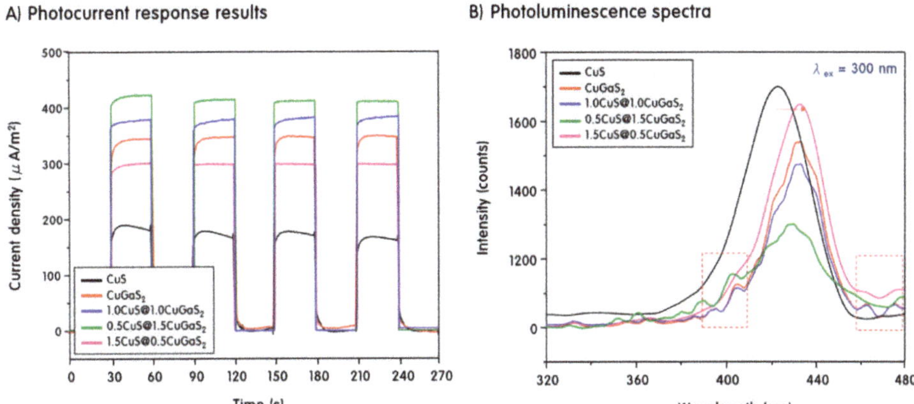

Figure 4. Photocurrent responses (**A**) and photoluminescence spectra (**B**) of CuS, $CuGaS_2$ and $CuS@CuGaS_2$ catalysts.

The transfer process of the photogenerated charge carrier is closely related to photoluminescence, and the photoluminescence measurement results are shown in Figure 4B. The position and intensity of the emission peak of photoluminescence (PL) differs depending on the kind of the catalyst and the vacancy [32]. Generally, as the intensity is increased, the recombination of the photogenerated electron and the hole pair is promoted and the photocatalytic activity is decreased [33]. When an excitation wavelength of 300 nm was irradiated, an emission peak was observed at about 420 nm in all of the CuS, $CuGaS_2$ and heterojunction $CuS@CuGaS_2$ nanoparticles. Especially, when Ga was added, the photoluminescence peak intensity of $CuGaS_2$ and the heterogeneous $CuS@CuGaS_2$ catalyst was slightly shifted to the long wavelength side. As the number of defect lattices generated in the Ga doping and heterojunction process increases, the electron trap site increases, which results in suppression of the recombination of photogenerated electrons and hole pairs. On the other hand, $CuGaS_2$ and heterojunction $CuS@CuGaS_2$ catalysts to which Ga was added, in comparison with pure CuS, had new emission peaks near 400 nm and 480 nm. According to Mehmood et al. [34], the blue emission peak at around 400 nm is due to a high energy defect due to the dopant. In addition, the electrons trapped by the dopant cannot generate excitons because they are bound by surface oxygen defects and other defects, thereby reducing the overall PL intensity. The blue-green emission peak at about 480 nm corresponds to the radiative transition of an electron to the deep donor level of the metal interstitials to an acceptor level of neutral V_{metal} [35]. In contrast to photocurrent measurement, the photoluminescence intensity decreased in the order of CuS > $1.5CuS@0.5CuGaS_2$ > $CuGaS_2$ > $1.0CuS@1.0CuGaS_2$ > $0.5CuS@1.5CuGaS_2$. As a result, the heterojunction $0.5CuS@1.5CuGaS_2$ catalyst showed the lowest photoluminescence intensity, and the Ga doping and heterojunction structure can play an important role in increasing the light efficiency by slowing the recombination between exciton and hole pairs the most.

Based on the results of photoluminescence measurements, further intensity modulated photovoltage spectroscopy (IMVS) measurements were performed to compare the excited electron recombination lifetime for CuS, $CuGaS_2$ and heterojunction $CuS@CuGaS_2$ catalysts, and the results are shown in Figure 5A. IMVS is a useful method to study lifetime of electrons, which relates to the electron recombination process [36]. The IMVS plot (A) of all samples showed a semicircular shape, and the larger the size of the semicircle in the IMVS results, the longer the recombination lifetime [37]. The electron recombination lifetime (B) was calculated from the IMVS plot using the equation [38] $sr = 1/2\ \pi f_{min}$, where f_{min} is the frequency of the minimum imaginary component of the plot. As with the results of the photoluminescence measurement, the electron recombination lifetime increased in the order of: $CuS < CuGaS_2 < CuS@CuGaS_2$. The increased IMVS electron lifetime indicates that the residence time at the electron trap site increases. Figure 5B shows that $CuGaS_2$ increases the recombination lifetime due to the effect of Ga dopant compared to CuS, but shows the longest recombination life when the two catalysts are hetero-bonded. These results indicate that defects or catalyst surfaces formed during the heterojunction process generate more trap sites, and that the electron recombination lifetime is accelerated while accelerating electron-transfer, which may exert an excellent photocatalytic activity.

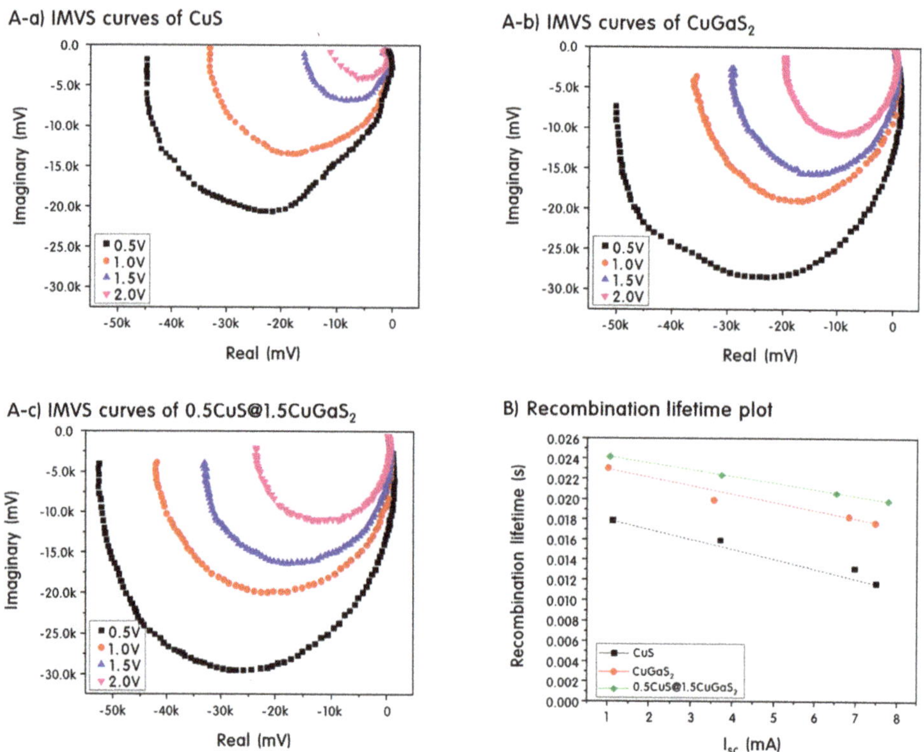

Figure 5. Intensity modulated photovoltage spectroscopy (IMVS) curves (**A**) and electron recombination lifetime (**B**) of CuS, $CuGaS_2$ and $CuS@CuGaS_2$ catalysts.

XPS was measured to investigate the chemical states of Cu, Ga and S ions in the surface state of the heterojunction $0.5CuS@1.5CuGaS_2$ sample. The results are shown in Figure 6. In Cu atomic spectra, peaks corresponding to Cu $2p_{3/2}$ and $2p_{1/2}$ were observed at 934.27 and 953.71 eV. A typical satellite peak of the Cu^{2+} oxidation state was observed at 942.78 and 963.06 eV, except for these two distinct peaks, indicating a defect in Cu^{2+}. This implies a vacancy in the surface state that occurs in vacancies or $0.5CuS@1.5CuGaS_2$ heterojunction processes in skeletal distortions due to ion size differences due to Ga addition [39]. In the Ga atomic spectrum, peaks corresponding to $2p_{3/2}$ and $2p_{1/2}$ were observed at 1118.87 and 1145.80 eV, which corresponds to the Ga-S bond. On the other hand, three fitting curves were separated in the S atomic spectrum. The peak at low binding energy (162.00 eV) corresponds to the Cu-S bond, and the peak at the high binding energy (165.28 eV) corresponds to the Ga-S bond [40]. Furthermore, the intermediate peak observed near 163.74 eV corresponds to sulfur vacancy, which is attributed to sulfur defects formed at the CuGaS$_2$ or heterojunction interface [41]. These XPS results demonstrate the presence of structural defects on the heterojunction 0.5CuS@1.5CuGaS$_2$ catalyst surface and predict that the site can act as a reactive site.

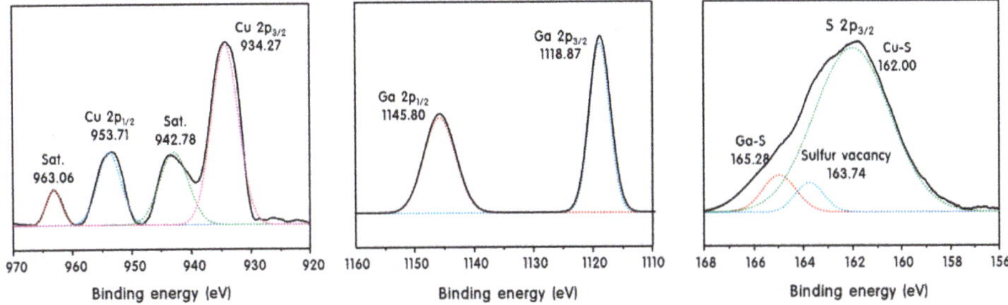

Figure 6. XPS spectra of heterojunction 0.5CuS@1.5CuGaS$_2$ catalysts.

Figure 7 summarizes the evolution of hydrogen from the photo splitting of aqueous methanol solution to CuS, CuGaS$_2$ and CuS@CuGaS$_2$ catalysts. Figure 7A shows the amount of hydrogen generated under 365 nm UV light source conditions of the catalyst. The pure CuS catalyst showed little hydrogen evolution even after 10 h of reaction. On the other hand, the amount of hydrogen generation in the CuGaS$_2$ catalyst was remarkably increased, and the amount of hydrogen produced after the reaction for 10 h reached 3000 μmol. In particular, the CuS@CuGaS$_2$ and 0.5CuS@1.5CuGaS$_2$ catalysts increased the amount of hydrogen generation further, reaching 3100 and 3250 μmol, respectively. Meanwhile, the heterojunction 1.5CuS@0.5CuGaS$_2$ catalyst with a high CuS ratio produced less hydrogen than CuGaS$_2$. This is considered to be due to the fact that the portion exposed on the surface of the catalyst has a large amount of CuS and is less influenced by CuGaS$_2$. According to the previous studies [42], the photo splitting process of methanol aqueous solution follows the following reaction:

$$CH_3OH + H_2O \rightarrow CO_2 + 6H^+ + 6e^- \tag{1}$$

$$6H^+ + 6e^- \rightarrow 3H_2 \tag{2}$$

$$CH_3OH + H_2O \rightarrow CO_2 + 3H_2 \tag{3}$$

According to the above photolytic decomposition method of methanol, hydrogen and carbon dioxide are produced, but in this study, carbon dioxide was not observed because it exists as a CO$_2$ ion in an aqueous solution. In addition, photolysis of methanol aqueous solution was further performed using a 150 W Xe lamp to confirm the catalytic activity in the visible region. The amount of hydrogen produced was reduced by about 1/20 compared to the UV light source, and the 0.5CuS@1.5CuGaS$_2$ catalyst, which was heterogeneously bonded to the CuGaS$_2$ catalyst, produced about 200 μmol of hydrogen. From these results, we have confirmed that CuGaS$_2$ and the heterogeneous CuS@CuGaS$_2$ catalyst exhibit optical activity even in the visible region, albeit in a small amount compared to UV light sources.

From these results, we proposed an improved photocatalytic decomposition of aqueous solution of CuS@CuGaS$_2$ catalyst as shown in Scheme 1. The valence band, conduction band, and band gap values of CuS [18] and CuGaS$_2$ [43] have already been reported in other studies and based on this, energy potential diagrams are shown together. In the heterojunction CuS@CuGaS$_2$, the band gap of CuGaS$_2$ contains CuS, and CuGaS$_2$ first acts as a light-absorbing agent. Electrons are excited from the valence band to the conduction band by light irradiation, and electrons generated from CuGaS$_2$ can move to the conduction band of the adjacent CuS. This transfer is thermodynamically favorable by band gap alignment, and the photogenerated electrons react with H^+ to produce H_2. At this time, S^{2-}/S_x^{2-}, which is a sacrifice material of the metal sulfide, captures holes by the following equation [44], and suppresses recombination between electron-hole pairs.

(1) $2\,S^{2-} + 2\,h^+ \rightarrow S_2^{2-}$

(2) $S^{2-} + 2h^+ \rightarrow S$

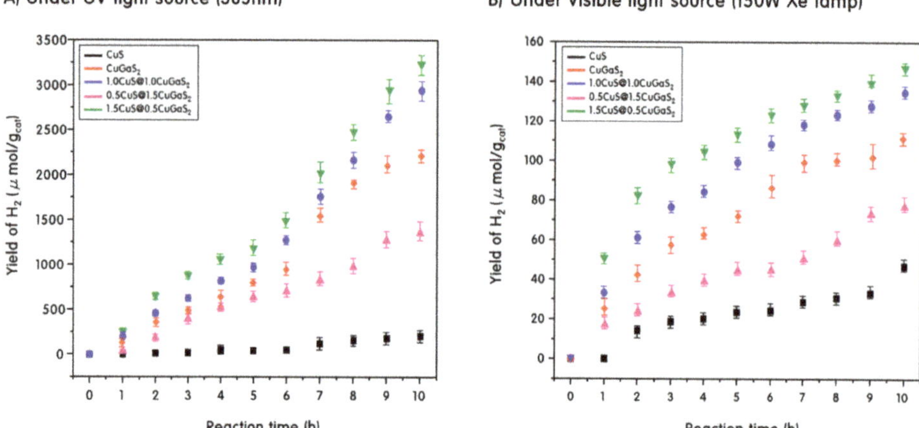

Figure 7. Evolution of H_2 for methanol aqueous solution photo-splitting under UV light source (**A**) and visible light source (**B**) for CuS, CuGaS$_2$ and CuS@CuGaS$_2$.

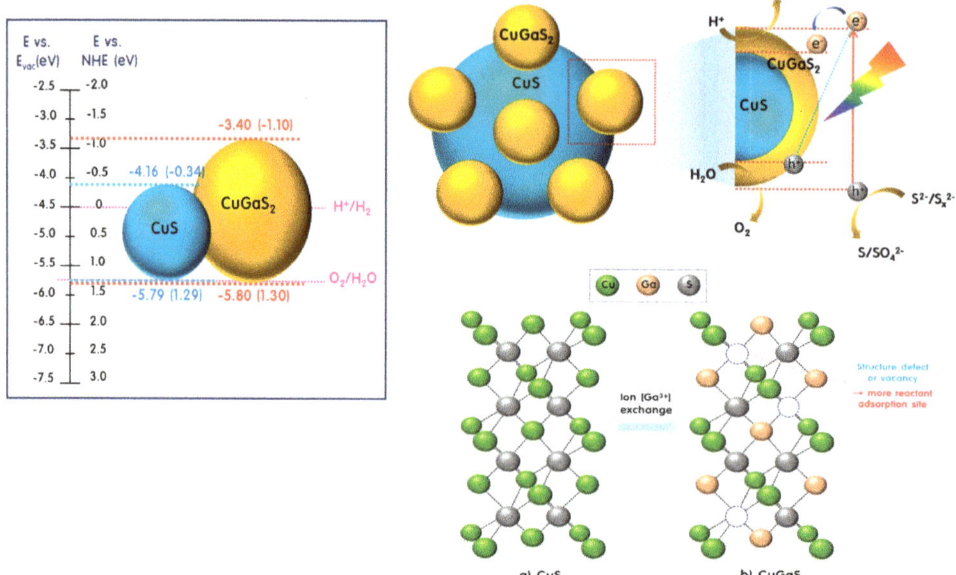

Scheme 1. The expected mechanism for methanol aqueous solution photo-splitting in photosystem with CuS@CuGaS$_2$ catalyst.

In addition, structural defects of the $CuGaS_2$ catalyst caused by the interface between the heterojunction $CuS@CuGaS_2$ catalyst and the $GaGaS_2$ catalyst may act as a trap site to accelerate hole transfer and prolong the electron lifetime. In $CuGaS_2$ structure, free electrons are gathered around Ga^{3+} for charge balance based on CuS structure, and sulfur vacancies due to structural defects are formed. These sites provide a place where the reactants can be adsorbed better, resulting in more products, and ultimately an improvement in photocatalytic activity. Moreover, structural defects at the $CuS@CuGaS_2$ catalyst interface form quasi-continuous energy levels and reduce ohmic contact to induce ohmic contact [28]. This contact recombines the holes formed in VB of CuS and the excited electrons in CB of $CuGaS_2$, which ultimately promotes the efficiency of photocatalytic activity by promoting the separation efficiency of the photogenerated charge pair in the heterojunction $CuS@CuGaS_2$ complex catalyst.

3. Experimental

3.1. Preparation of CuS, CuGaS₂ and CuS@CuGaS₂ Nanoparticles

CuS and $CuGaS_2$ nanoparticles were prepared using a typical sol-gel synthesis method [45]. Copper (II) nitrate trihydrate ($Cu(NO_3)_2 \cdot 3H_2O$, 99.0%, Junsei Chemical, Tokyo, Japan), Gallium (III) nitrate hydrate ($Ga(NO_3)_3 \cdot xH_2O$, 99.9%, Alfa Aesar, Tewksbury, MA, USA) and Thiourea (CH_4N_2S, 98.0%, Junsei Chemical, Tokyo, Japan) were used as starting materials for Cu, Ga and S, respectively. First, in order to synthesize CuS, $Cu(NO_3)_2 \cdot 3H_2O$ and CH_4N_2S were dissolved in ethylene glycol at a molar ratio of 1:2, mixed well and aged at 180 °C for 8 h. The resulting powder was treated at 400 °C for 4 h under an argon atmosphere to obtain black CuS nanoparticles. In the synthesis of $CuGaS_2$ particles, only the molar ratio of $Cu(NO_3)_2 \cdot 3H_2O$, $Ga(NO_3)_3 \cdot xH_2O$ and CH_4N_2S was changed to 1:1.25:4 during the synthesis of CuS, respectively.

On the other hand, heterogeneous $CuS@CuGaS_2$ nanoparticles were obtained by the impregnation method [46] using prepared CuS and $CuGaS_2$. The synthesis procedure was as follows. The amounts of CuS and $CuGaS_2$ added were different. CuS was added to ethanol, and the mixture was stirred for 2 h, then $CuGaS_2$ was added and stirred sufficiently. The homogeneously stirred solution was separated into powdery samples by centrifugation and dried at 80 °C for 24 h. Thereafter, the resultant was again annealed at 200 °C for 2 h in order to remove impurities and increase the bonding strength, finally obtaining a heterogeneous $CuS@CuGaS_2$ catalyst.

3.2. Characterization of CuS, CuGaS₂ and CuS@CuGaS₂ Nanoparticles

X-ray diffraction (XRD, MPD, PANalytical, Almelo, The Netherlands) was used to analyze the crystal structure of the prepared CuS, $CuGaS_2$ and heterojuntion $CuS@CuGaS_2$ nanoparticles. XRD was measured at a 2θ angle of 20–100° using nickel-filtered CuKα ($\lambda = 1.5056$ Å) radiation (40 kV, 30 mA). The shape and size of the particles were confirmed using high-resolution transmission electron microscopy (TEM, H-7600, Hitachi, Tokyo, Japan) and scanning electron microscopy (SEM, S-4100, Hitachi, Tokyo, Japan). In addition, energy-dispersive X-ray spectroscope (EDS, EX-250, Horiba, Kyoto, Japan) analysis was used to identify the atomic composition of CuS, $CuGaS_2$, and heterogeneous $CuS@CuGaS_2$ nanoparticles.

The diffuse reflection spectra of the particles were obtained using a UV-Vis spectrophotometer (Neosys-2000, SCINCO, Daejeon, Korea). The recombination tendency between the photogenerated electron-hole pair (e^-/h^+) of the catalyst was determined using a photoluminescence spectroscopy (PL, FS-2, SCINCO, Daejeon, Korea) equipped with a 150 W continuous Xenon lamp light source.

In addition, photocurrent and intensity modulated photovoltage spectroscopy (IMVS) measurements were taken with a two-electrode system to confirm the behavior of the photogenerated charge carrier. The catalyst was coated on fluorine doped tin oxide (FTO) glass to form a cell, and a platinum wire was used as a counter electrode. The catalyst coated on the FTO glass with a certain unit area was used as the working electrode and photocurrent was measured by irradiating light at intervals of 30 s. The IMVS measurement was also performed using a visible light source in a two-electrode system, and the recombination lifetime of electrons was confirmed through the measurement.

3.3. Hydrogen Production by Photo Splitting of Methanol Aqueous Solution Using CuS, CuGaS$_2$ and Heterojunction CuS@CuGaS$_2$ Catalyst

The photocatalytic decomposition of methanol aqueous solution was carried out using a liquid photoreactor prepared in our laboratory, which was reported in previous work [47]. First, the photocatalytic decomposition of methanol aqueous solution using a UV light source was performed using a pyrex reactor. 1.0 L of a mixed solution of 500 mL of methanol and 500 mL of distilled water was put into the reactor, and 0.5 g of the synthesized CuS, CuGaS$_2$ and CuS@CuGaS$_2$ powder was added. Light was irradiated using a UV-lamp (3×6 cm^{-2} = 18 W cm^{-2}, length 30 cm, diameter 2.0 cm, Shinan, Pochon, Korea) at a wavelength of 365 nm and the reaction was performed for a total of 10 h. The photocatalytic decomposition of methanol aqueous solution using a visible light source was performed in a quartz reactor using a 150 W Xe lamp. The resulting gas was analyzed by gas chromatography (GC, DS7200, Donam Company, Gwangju, Korea) equipped with a thermal conductivity detector (TCD). The following GC conditions were used: TCD detector, Carboxen-1000 column (Bruker, Billerica, MA, USA), and the injection, oven and detector temperatures of 423, 393 and 473 K, respectively.

4. Conclusions

We have synthesized CuS, CuGaS$_2$, and heterojunction CuS@CuGaS$_2$ catalysts for hydrogen production through methanol aqueous photo splitting. The interface between the heterojunction CuS@CuGaS$_2$ catalyst and the structural defect of CuGaS$_2$ formed by the addition of Ga^{3+} to CuS acted as a trap site. This trap site accelerates the electron-transfer, indicating a high photocurrent density value in the photocurrent results and excellent charge separation efficiency. In addition, compared to pure CuS and CuGaS$_2$, as shown by photoluminescence and IMVS measurements, recombination between the excited electron-hole pairs in the heterojunction catalyst was suppressed, resulting in higher electron lifetime. As a result, the heterojunction CuS@CuGaS$_2$ catalyst produced a significant amount of hydrogen gas, up to 3250 and 200 μmol, through photo splitting of aqueous methanol solution under UV and visible light irradiation, showing a significant increase in photocatalytic activity.

Author Contributions: Conceptualization, M.K.; Data curation, N.S. and J.Y.D.; Formal analysis, N.S. and J.Y.D.; Investigation, J.N.H. and Y.K.; Methodology, J.N.H. and Y.-S.Y.; Supervision, M.K.; Writing—original draft, J.Y.D.; Writing—review & editing, M.K.

Funding: This work was supported by the National Research Foundation of Korea (NRF) grant funded by the Korean government (MSIT) (No. 2018R1A2B6004746).

Conflicts of Interest: The authors declare no conflict of interest.

References

1. Stam, W.; Gradmann, S.; Altantzis, T.; Ke, X.; Baldus, M.; Bals, S.; Donega, C. Shape Control of Colloidal Cu$_{2-x}$S Polyhedral Nanocrystals by Tuning the Nucleation Rates. *Chem. Mater.* **2016**, *28*, 6705–6715. [CrossRef] [PubMed]

2. Zhao, Y.; Burda, C. Development of plasmonic semiconductor nanomaterials with copper chalcogenides for a future with sustainable energy materials. *Energy Environ. Sci.* **2012**, *5*, 5564–5577. [CrossRef]

3. Sun, S.; Li, P.; Liang, S.; Yang, Z. Diversified copper sulfide (Cu$_{2-x}$S) micro /nanostructures: A comprehensive review on synthesis, modifications and applications. *Nanoscale* **2017**, *9*, 11357–11404. [CrossRef] [PubMed]

4. Tian, Q.; Hu, J.; Zhu, Y.; Zou, R.; Chen, Z.; Yang, S.; Li, R.; Su, Q.; Han, Y.; Liu, X. Sub-10 nm Fe$_3$O$_4$@Cu$_{2-x}$S Core−Shell Nanoparticles for Dual-Modal Imaging and Photothermal Therapy. *J. Am. Chem. Soc.* **2013**, *135*, 8571–8577. [CrossRef] [PubMed]

5. Abdelhady, A.L.; Ramasamy, K.; Malik, M.A.; O'Brien, P.; Haigh, S.J.; Raftery, J. New routes to copper sulfide nanostructures and thin films. *J. Mater. Chem.* **2011**, *21*, 17888–17895. [CrossRef]

6. Wang, N.; Gao, C.; Han, Y.; Huang, X.; Xu, Y.; Cao, X. Detection of human immunoglobulin G by label-free electrochemical immunoassay modified with ultralong CuS nanowires. *J. Mater. Chem. B* **2015**, *3*, 3254–3259. [CrossRef]

7. Li, Z.; Yang, H.; Ding, Y.; Xiong, Y.; Xie, Y. Solution-phase template approach for the synthesis of Cu_2S nanoribbons. *Dalton Trans.* **2006**, 149–151. [CrossRef] [PubMed]

8. Yan, C.; Rosei, F. Hollow micro/nanostructured materials prepared by ion exchange synthesis and their potential applications. *New J. Chem.* **2014**, *38*, 1883–1904. [CrossRef]

9. Li, D.; Zheng, Z.; Lei, Y.; Ge, S.; Zhang, Y.; Zhang, Y.; Wong, K.W.; Yang, R.; Lau, W.M. Design and growth of dendritic Cu_2-xSe and bunchy CuSe hierarchical crystalline aggregations. *CrystEngComm* **2010**, *12*, 1856–1861. [CrossRef]

10. Jayasekharan, T.; Gupta, S.L.; Dhiman, V. Binding of Cu^+ and Cu^{2+} with peptides: Peptides = oxytocin, Arg8-vasopressin, bradykinin, angiotensin-I, substance-P, somatostatin, and neurotensin. *J. Mass Spectrom.* **2018**, *53*, 296–313. [CrossRef]

11. Sant, S.B. *Nanoparticles: From Theory to Applications, Gunter Schmid*, 2nd ed.; Wiley-VCH Verlag: Weinheim, Germany, 2010; ISBN 978-3-527-32589-4.

12. Saranya, M.; Santhosh, C.; Ramachandran, R.; Kollu, P.; Saravanan, P.; Vinoba, M.; Jeong, S.K.; Grace, A.N. Hydrothermal growth of CuS nanostructures and its photocatalytic properties. *Powder Technol.* **2014**, *252*, 25–32. [CrossRef]

13. Zhang, J.; Yu, J.; Zhang, Y.; Li, Q.; Gong, J.R. Visible light photocatalytic H_2-production activity of CuS/ZnS porous nanosheets based on photoinduced interfacial charge transfer. *Nano Lett.* **2011**, *11*, 4774–4779. [CrossRef] [PubMed]

14. Lee, H.; Kwak, B.S.; Park, N.-K.; Baek, J.-I.; Ryu, H.-J.; Kang, M. Assembly of a check-patterned CuSx-TiO$_2$ film with an electron-rich pool and its application for the photoreduction of carbon dioxide to methane. *Appl. Surf. Sci.* **2017**, *393*, 385–396. [CrossRef]

15. Salak, A.N.; Ignatenko, O.V.; Zhaludkevich, A.L.; Lisenkov, A.D.; Starykeyich, M.; Zheludkevich, M.; Ferreira, M.G.S. High-pressure zinc oxysulphide phases in the ZnO–ZnS system. *Phys. Status Solidi* **2015**, *212*, 791–795. [CrossRef]

16. Yue, M.; Wang, R.; Ma, B.; Cong, R.; Gao, W.; Yang, T. Superior performance of CuInS$_2$ for photocatalytic water treatment: Full conversion of highly stable nitrate ions into harmless N_2 under visible light. *Catal. Sci. Technol.* **2016**, *6*, 8300–8309. [CrossRef]

17. Han, F.; Kambala, V.S.R.; Srinivasan, M.; Jajarathnam, D.; Naidu, R. Tailored titanium dioxide photocatalysts for the degradation of organic dyes in wastewater treatment: A review. *Appl. Catal. A* **2009**, *359*, 25–40. [CrossRef]

18. Do, J.Y.; Chava, R.K.; Kim, S.K.; Nahm, K.; Park, N.-K.; Hong, J.-P.; Lee, S.J.; Kang, M. Fabrication of core@interface:shell structured CuS@CuInS$_2$:In$_2$S3 particles for highly efficient solar hydrogen production. *Appl. Surf. Sci.* **2018**, *451*, 86–98. [CrossRef]

19. Chae, J.; Lee, J.; Jeong, J.H.; Kang, M. Hydrogen Production from Photo Splitting of Water Using the Ga-incorporated TiO$_2$s Prepared by a Solvothermal Method and Their Characteristics. *Bull. Korean Chem. Soc.* **2009**, *30*, 302–309.

20. Ravi, S.; Gopi, C.V.V.M.; Kim, H. Enhanced electrochemical capacitance of polyimidazole coated covellite CuS dispersed CNT composite materials for application in supercapacitors. *Dalton Trans.* **2016**, *45*, 12362–12372. [CrossRef]

21. Kundu, J.; Pradhan, D. Influence of precursor concentration, surfactant and temperature on the hydrothermal synthesis of CuS: Structural, thermal and optical properties. *New J. Chem.* **2013**, *37*, 1470–1479. [CrossRef]

22. Liu, S.; Nie, L.; Yuan, R. Growth, structure and optical characterization of CuGaS$_2$ thin films obtained by spray pyrolysis. *Chalcogenide Lett.* **2015**, *12*, 111–116.

23. Kimi, M.; Yuliati, L.; Shamsuddin, M. Preparation of High Activity Ga and Cu Doped ZnS by Hydrothermal Method for Hydrogen Production under Visible Light Irradiation. *J. Nanomater.* **2015**, *2015*, 200. [CrossRef]

24. Scrivener, K.L.; Fullmann, T.; Galucci, E.; Walenta, G.; Bermejo, E. Quantitative study of Portland cement hydration by X-ray diffraction/Rietveld analysis and independent methods. *Cem. Concr. Res.* **2004**, *34*, 1541–1547. [CrossRef]

25. Do, J.Y.; Lee, J.H.; Park, N.-K.; Lee, T.J.; Lee, S.T.; Kang, M. Synthesis and characterization of $Ni_{2-x}Pd_xMnO_4/\gamma\text{-}Al_2O_3$ catalysts for hydrogen production via propane steam reforming. *Chem. Eng. J.* **2018**, *334*, 1668–1678. [CrossRef]

26. Lin, H.; Huang, C.P.; Li, W.; Ni, C.; Shah, I.; Tseng, Y.-H. Size dependency of nanocrystalline TiO_2 on its optical property and photocatalytic reactivity exemplified by 2-chlorophenol. *Appl. Catal. B.* **2006**, *68*, 1–11. [CrossRef]

27. Huxter, V.M.; Mirkovic, T.; Nair, S.; Scholes, G.D. Demonstration of bulk semiconductor optical properties in processable Ag_2S and EuS nanocrystalline systems. *Adv. Mater.* **2008**, *20*, 2439–2443. [CrossRef]

28. Maeda, T.; Yu, Y.; Chen, Q.; Ueda, K.; Wada, T. Crystallographic and optical properties and band diagrams of $CuGaS_2$ and $CuGa_5S_8$ phases in Cu-poor $Cu_2S\text{-}Ga_2S_3$ pseudo-binary system. *Jpn. J. Appl. Phys.* **2017**, *56*, 04CS12. [CrossRef]

29. Aliaga, J.; Cifuentes, N.; González, G.; Sotomayor-Torres, C.; Benavente, E. Enhancement Photocatalytic Activity of the Heterojunction of Two-Dimensional Hybrid Semiconductors ZnO/V_2O_5. *Catalysts* **2018**, *8*, 374. [CrossRef]

30. Tessler, N.; Rappaport, N. Excitation density dependence of photocurrent efficiency in low mobility semiconductors. *J. Appl. Phys.* **2004**, *96*, 1083–1087. [CrossRef]

31. Xiang, Q.; Yu, J.; Jaroniec, M. Enhanced photocatalytic H_2-production activity of graphene-modified titania nanosheets. *Nanoscale* **2011**, *3*, 3670–3678. [CrossRef]

32. Djurišić, A.B.; Leung, Y.H.; Tam, K.H.; Ding, L.; Ge, W.K.; Chen, H.Y.; Gwo, S. Green, yellow, and orange defect emission from ZnO nanostructures: Influence of excitation wavelength. *Appl. Phys. Lett.* **2006**, *88*, 103107–103110. [CrossRef]

33. Yu, J.C.; Yu, J.; Ho, W.; Jiang, Z.; Zhang, L. Effects of F^- Doping on the Photocatalytic Activity and Microstructures of Nanocrystalline TiO_2 Powders. *Chem. Mater.* **2002**, *14*, 3808–3816. [CrossRef]

34. Mehmood, F.; Iqbal, J.; Ismail, M.; Mehmood, A. Ni doped WO_3 nanoplates: An excellent photocatalyst and novel nanomaterial for enhanced anticancer activities. *J. Alloy. Compd.* **2018**, *746*, 729–738. [CrossRef]

35. Xie, Z.; Sui, Y.; Buckeridge, J.; Catlow, C.R.A.; Keal, T.W.; Sherwood, P.; Walsh, A.; Farrow, M.R.; Scanlon, D.O.; Woodley, S.M.; et al. Donor and Acceptor characteristics of Native Point Defects in GaN. *arXiv*, 2018; arXiv:1803.06273.

36. Cen, J.; Wu, Q.; Liu, M.; Orlov, A. Devloping new understanding of photo-electrochemical water splitting via in-situ techniques: A review on recent progress. *Green Energy Environ.* **2017**, *2*, 100–111. [CrossRef]

37. Yang, Y.; Ri, K.; Mei, A.; Liu, L.; Hu, M.; Liu, T.; Li, X.; Han, H. The size effect of TiO_2 Nanoparticles on a printable mesoscopic perovskite solar cell. *J. Mater. Chem. A* **2015**, *3*, 9103–9107. [CrossRef]

38. Bavarian, M.; Nejati, S.; Lau, K.K.S.; Lee, D.; Soroush, M. Theoretical and Experimental study of a Dye-Sensitized Solar Cell. *Ind. Eng. Chem. Res.* **2014**, *53*, 5234–5247. [CrossRef]

39. Ludwing, J.; An, L.; Pattengale, B.; Kong, Q.; Zhang, X.; Xi, P.; Huang, J. Ultrafast hole trapping and relaxation dynamics in p-type CuS nanodisks. *J. Phys. Chem. Lett.* **2015**, *6*, 2671–2675. [CrossRef] [PubMed]

40. Song, Y.; Li, Y.; Zhu, L.; Pan, Z.; Jiang, Y.; Wang, P.; Zhou, Y.-N.; Fang, F.; Hu, L.; Sun, D. $CuGaS_2$ nanoplates: A robust and self-healing anode for Li/Na ion batteries in a wide temperature range of 268–318 K. *J. Mater. Chem. A* **2018**, *6*, 10886–11093. [CrossRef]

41. Dong, J.; Zeng, X.; Xia, W.; Zhang, X.; Zhou, M.; Wang, C. Ferromagnetic behavior of non-stoichiometric ZnS microspheres with a nanoplate-netted surface. *RSC Adv.* **2017**, *7*, 20874–20881. [CrossRef]

42. Do, J.Y.; Choi, S.; Nahm, K.; Kim, S.K.; Kang, M. Reliable hydrogen production from methanol photolysis in aqueous solution by a harmony between In and Zn in bimetallic zinc indium sulfide. *Mater. Res. Bull.* **2018**, *100*, 234–242. [CrossRef]

43. Cui, H.; Li, B.; Li, Z.; Li, X.; Xu, S. Z-scheme based $CdS/CdWO_4$ heterojunction visible light photocatalyst for dye degradation and hydrogen evolution. *Appl. Surf. Sci.* **2018**, *455*, 831–840. [CrossRef]

44. Kraeutler, B.; Bard, A.J. Heterogeneous Photocatalytic Decomposition of Saturated Carboxylic Acids on TiO_2 Powder. Decarboxylative Route to Alkanes. *J. Am. Chem. Soc.* **1978**, *100*, 5985–5992. [CrossRef]

45. Peña-Garcia, R.; Guerra, Y.; Farias, B.V.M.; Martínez Buitrago, D.; Franco, A., Jr.; Padrón-Hernández, E. Effects of temperature and atomic disorder on the magnetic phase transitions in ZnO nanoparticles obtained by sol-gel method. *Mater. Lett.* **2018**, *233*, 146–148.

46. Zhang, R.; Xu, J.; Lu, C.; Xu, Z. Photothermal application of SmCoO$_3$/SBA-15 catalysts synthesized by impregnation method. *Mater. Lett.* **2018**, *228*, 199–202. [CrossRef]
47. Son, N.; Do, J.Y.; Kang, M. Characterization of core@shell-structured ZnO@Sb$_2$S$_3$ particles for effective hydrogen production from water photo spitting. *Ceram. Int.* **2017**, *43*, 11250–11259. [CrossRef]

 *catalysts*

Article

In-Situ Synthesis of Nb_2O_5/g-C_3N_4 Heterostructures as Highly Efficient Photocatalysts for Molecular H_2 Evolution under Solar Illumination

Faryal Idrees [1,2,3,*], **Ralf Dillert** [1,2], **Detlef Bahnemann** [1,2,*], **Faheem K. Butt** [4] **and Muhammad Tahir** [3]

1   Institut für Technische Chemie, Gottfried Wilhelm Leibniz Universität Hannover, Callinstrasse 3, D-30167 Hannover, Germany; dillert@iftc.uni-hannover.de
2   Laboratorium für Nano- und Quantenengineering, Gottfried Wilhelm Leibniz Universität Hannover, Schneiderberg 39, D-30167 Hannover, Germany
3   Department of Physics, The University of Lahore, 1-Km Raiwind, Lahore 53700, Pakistan; tahir94@gmail.com
4   Department of Physics, Division of Science and Technology, University of Education, College Road, Township, Lahore 54770, Pakistan; faheem.khurshid@phys.uol.edu.pk
*   Correspondence: idrees@iftc.uni-hannover.de (F.I.); bahnemann@iftc.uni-hannover.de (D.B.)

Received: 7 January 2019; Accepted: 7 February 2019; Published: 11 February 2019

Abstract: This work focuses on the synthesis of heterostructures with compatible band positions and a favourable surface area for the efficient photocatalytic production of molecular hydrogen (H_2). In particular, 3-dimensional Nb_2O_5/g-C_3N_4 heterostructures with suitable band positions and high surface area have been synthesized employing a hydrothermal method. The combination of a Nb_2O_5 with a low charge carrier recombination rate and a g-C_3N_4 exhibiting high visible light absorption resulted in remarkable photocatalytic activity under simulated solar irradiation in the presence of various hole scavengers (triethanolamine (TEOA) and methanol). The following aspects of the novel material have been studied systematically: the influence of different molar ratios of Nb_2O_5 to g-C_3N_4 on the heterostructure properties, the role of the employed hole scavengers, and the impact of the co-catalyst and the charge carrier densities affecting the band alignment. The separation/transfer efficiency of the photogenerated electron-hole pairs is found to increase significantly as compared to that of pure Nb_2O_5 and g-C_3N_4, respectively, with the highest molecular H_2 production of 110 mmol/g·h being obtained for 10 wt % of g-C_3N_4 over Nb_2O_5 as compared with that of g-C_3N_4 (33.46 mmol/g·h) and Nb_2O_5 (41.20 mmol/g·h). This enhanced photocatalytic activity is attributed to a sufficient interfacial interaction thus favouring the fast photogeneration of electron-hole pairs at the Nb_2O_5/g-C_3N_4 interface through a direct Z-scheme.

Keywords: Niobium(V) oxide; graphitic carbon nitride; hydrothermal synthesis; H_2 evolution; photocatalysis; heterostructures; Z-Scheme

---

1. Introduction

Renewable energy sources are currently needed by our society to address the foreseable future energy crisis and growing environmental issues. The production of molecular H_2 through photoelectrochemical or photocatalytic water splitting is a viable replacement of fossil fuels [1]. In the past, due to certain limitations of the most frequently employed photocatalyst, TiO_2, many other photocatalysts have been developed and explored for photocatalytic molecular H_2 evolution. Their significant limitations are the high recombination rate of photogenerated electron-hole pairs and unfavorable band edges, hence their low photocatalytic activity. For favorable band edges, CuO and Cu_2O are thought of as good alternatives but their stability and effective light absorption present additional issues. Somehow, these issues have been solved by adopting the atomic layer deposition (ALD) technique and doping, etc. [2–4]. The non-toxicity, facile synthesis, high visible light absorption and good physicochemical stability of g-C_3N_4 have made it an amiable photocatalyst [5–9]. However, the fast recombination rate of photogenerated charge carriers has reduced its charismatic effect. Some researchers proposed to solve this problem by forming heterojunctions with TiO_2 [10], WS_2 [11], $BiOCl$ [12] and WO_3 [13,14]. Still, there is considerable scope to develop other photocatalysts, with the particular focus being on facile synthesis, low toxicity, easy accessibility, and, most importantly, high photocatalytic activity [15].

Niobium pentoxide (Nb_2O_5) is an n-type and wide bandgap (3.4 eV) semiconductor, which has been extensively investigated in recent years for electrochemical, photocatalytic, and energy storage applications [16,17]. Its light absorption can be effectively shifted into the visible region by synthesizing composites with small bandgap materials. Noticeably, heterojunctions prepared with small amounts of Nb_2O_5 have shown a significant improvement in the photocatalytic activity of TiO_2 [8], ZnO [18], and g-C_3N_4 [7,9,19]. To date, there are only a few reports regarding the synthesis of g-C_3N_4/Nb_2O_5 heterojunctions and their photocatalytic properties. Y. Z. Hong et al. [19] and Q. Z. Huang et al. [20] have prepared g-C_3N_4/Nb_2O_5 heterojunctions facing, however, considerable limitations concerning the control of the size and shape of Nb_2O_5, and resulting in a rather small specific surface area. Moreover, to the best of our knowledge, Nb_2O_5/g-C_3N_4 heterostructures have so far not been explored for molecular H_2 evolution.

To increase photocatalytic activities, semiconductor-semiconductor heterostructures exhibiting lower recombination rates of photogenerated electron-hole pairs have been studied. Depending on the energetic situation, these heterostructures have been classified as type-II heterostructures and Z-scheme heterostructures, respectively (Figure 1; for a detailed discussion see the Supporting Information) [20–22]. The direct Z-scheme system seems promising for overcoming the limitations associated with enhanced photocatalytic activity, due to the strong oxidation and reduction potential developed at different active sites [21].

As depicted in Figure 1, a conventional type-II heterostructure can easily be converted into a Z-scheme structure by controlling the Fermi level or the band potentials. Moreover, the Fermi level and band potentials can also be modulated to attain the Z-scheme by the addition of suitable hole scavengers such as triethanolamine (TEOA), which has been reported to exhibit a larger H_2 evolution rate than methanol [19,22].

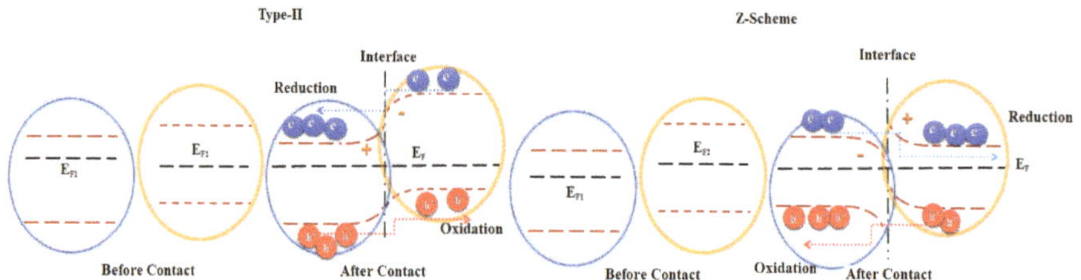

Figure 1. Schematic illustration of Type-II and Z-Scheme systems and their interfacial band bending under bandgap irradiation.

In the present work, Nb_2O_5/g-C_3N_4 Z-scheme heterojunctions were prepared considering their suitable band edges for photocatalytic H_2 production, i.e., Nb_2O_5 ($E_{CB} = -0.69$ V, $E_{VB} = 2.32$ V) and g-C_3N_4 ($E_{CB} = -1.68$ V, $E_{VB} = 0.88$ V) vs. NHE (pH = 7). In comparison to previous reports, we obtained a controlled shape of the heterojunction with a specific surface area as high as 227 m^2 g^{-1}. The prepared heterojunctions were tested for H_2 evolution without and with deposited platinum (Pt) acting as a cocatalyst, thus faciliating the interfacial electron transfer. The role of hole scavengers in the overall mechanism of the hydrogen evolution reaction was also investigated. A significant increase in the H_2 evolution rate was observed in the presence of TEOA compared with methanol. These findings have suggested a possible change in the photocatalytic mechanism to increase the evolution rate to such an extent. An excellent H_2 production rate of 110 mmol/h·g was found by employing platinized Nb_2O_5/g-C_3N_4 as the photocatalyst and TEOA as the hole scavenger. Noticeably, the high molecular H_2 rate could also be associated with the tuning of shape, size and structures. The compact interfacial development between Nb_2O_5/g-C_3N_4 also suggested that a Z-scheme system was formed, thus facilitating fast charge carrier separation and excellent photocatalytic performance as compared with simple physical mixing. The effect of different ratios of g-C_3N_4 to Nb_2O_5 has also been studied in detail.

2. Results and Discussion

2.1. Synthesis Procedure

Nb_2O_5 (NBO) and Nb_2O_5/g-C_3N_4 (NBCN) heterostructures were synthesized via a hydrothermal synthesis considering its possible principal advantages such as: (a) attaining porous structures with high surface areas; (b) reagents mixing at the atomic level, and (c) high reaction rates at a low reaction temperature due to the atomic mixing level [23–25]. Thus, a suitable strategy to tune the shape, size, and structure of Nb_2O_5 (NBO) and Nb_2O_5/g-C_3N_4 (NBCN) heterostructures with a high specific surface area and a sufficient contact interface was employed here.

NBO was synthesized via the oxidant-peroxo method (OPM). A niobium salt ($NbCl_5$) was dissolved in diluted nitric acid (HNO_3) to avoid salt residuals. In a second step H_2O_2 was added to the prepared solution to remove chloride ions by an oxidation–reduction process. The resulting yellow solution (pH = 0.5) indicated the presence of the water-soluble niobium peroxo-complex $[Nb(O_2)_4]^{3-}$ (named as NPC). A possible reaction is provided in Figure 2. The decomposition of H_2O_2 into molecular oxygen possibly accelerated the condensation reaction. Amorphous hierarchical spheres of Nb_2O_5 and $Nb_2O_5 \cdot nH_2O$ were obtained due to an excess amount of H_2O_2. Finally, annealing in the 200–500 °C range resulted in the formation of Nb_2O_5 with adequate composition and controlled morphology.

$$2NbCl_5 + 5n\,H_2O \longrightarrow Nb_2O_5 \cdot nH_2O + 10\,HCl$$

$$3H_2O_2 + 2HCl \longrightarrow 4\,H_2O + O_2 + Cl_2$$

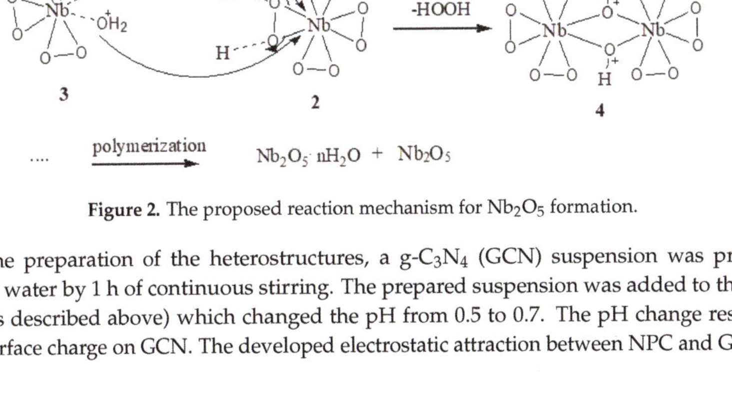

Figure 2. The proposed reaction mechanism for Nb_2O_5 formation.

For the preparation of the heterostructures, a g-C_3N_4 (GCN) suspension was prepared in de-ionised water by 1 h of continuous stirring. The prepared suspension was added to the solution of NPC (as described above) which changed the pH from 0.5 to 0.7. The pH change resulted in a positive surface charge on GCN. The developed electrostatic attraction between NPC and GCN favors the effective in-situ formation of heterojunctions with a controlled morphology [26]. This procedure was employed to prepare other heterostructures with a different wt % of GCN (as described in the experimental sections). Heterostructures have also been prepared by physical mixing of NBO and g-C$_3$N$_4$ (the method is described in the supporting information). A schematic presentation of the synthetic procedure is given in Figure 3.

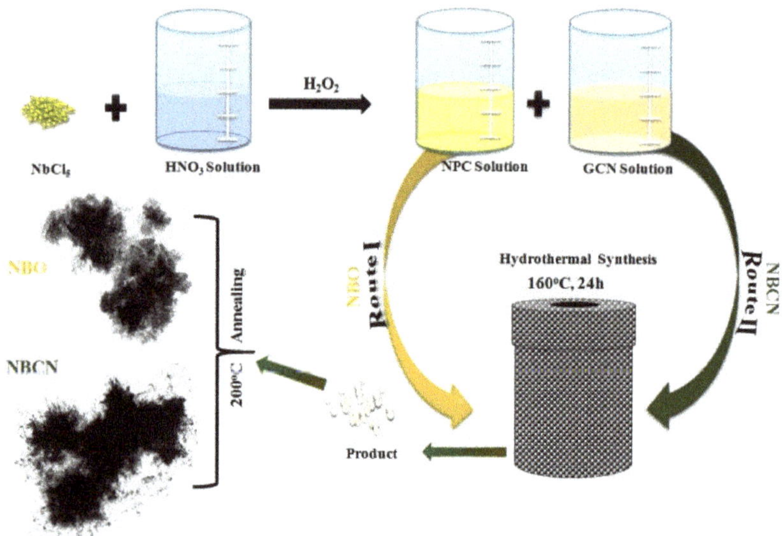

Figure 3. Schematic illustration of the hydrothermal routes followed for Nb$_2$O$_5$ (NBO) and Nb$_2$O$_5$/g-C$_3$N$_4$ (NBCN) synthesis.

2.2. Structural and Morphological Characterization

The phases and the crystal structures of as-prepared NBO, GCN, and NBCN-X were analyzed by XRD (Figure 4). The XRD data obtained for a NBO sample before annealing corresponded to a mixture of niobic acid and amorphous NBO. By annealing at 200 °C the mixture converted to the amorphous pseudohexagonal phase of Nb$_2$O$_5$ (TT-NBO, JCPDS#30-0873). No significant change in the phase was observed for samples annealed at temperatures between 200 °C and 400 °C. However, annealing at 500 °C results in the transformation of the amorphous pseudohexagonal phase of NBO in to the pseudohexagonal phase (JCPDS#30-0873) of NBO. The corresponding XRD results are provided in the supporting information in Figure S1. In the XRD of GCN, two peaks are observed at 13.4° and 27.0°, which were associated with the (100) and (002) diffraction planes, respectively. The (100) distinct peak was due to the characteristic inter-planar staking of aromatic systems, while the (002) peak corresponded to the inter-layer structural packing [26]. The XRDs of NBCN-X (X = 1–4) exhibited combinatory characteristic peaks of GCN and NBO. The characteristic peak of GCN (002) become stronger with an increasing amount of GCN, indicating a significant coupling between the GCN and NBO.

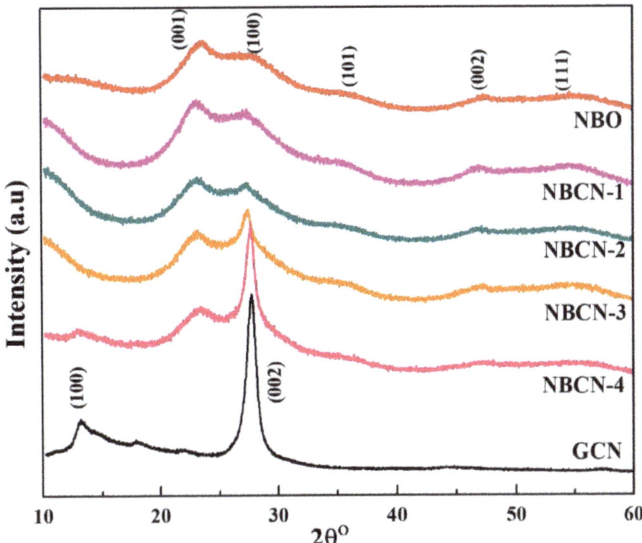

Figure 4. XRD patterns of NBO, g-C$_3$N$_4$ (GCN), NBCN-1, NBCN-2, NBCN-3, and NBCN-4.

The FTIR patterns further confirmed the development of compact heterostructure interfaces, as shown in Figure 5a. The amorphous pseudohexagonal phase of NBO possessed slightly distorted NbO$_6$, NbO$_7$, and NbO$_8$ groups and some highly distorted NbO$_6$ sites (a scheme is provided in Figure 5b). The peak at 665 cm^{-1} has been assigned to the symmetric stretching mode of Nb–O polyhedra (NbO$_6$, NbO$_7$, and NbO$_8$) [27]. The broad shoulder between 850 and 970 cm^{-1} has been assigned to the stretching of Nb=O groups. The removal of coordinated water further distorted the highly distorted NbO$_6$ octahedra, which led to the formation of Nb=O bonds [28]. The peaks at 414 cm^{-1} and 459 cm^{-1} have been attributed to the symmetric stretching v_s[Nb(O)$_2$] and the asymmetric stretching v_a[Nb(O)$_2$], respectively. Their occurrence indicated the presence of a small amount of coordinated peroxide on the Nb(V) ions [29]. With increasing calcination temperatures, the peak at 459 cm^{-1} first decreased from 200–400 °C and then disappeared entirely at 500 °C. The peaks at 3753 cm^{-1} and 1559 cm^{-1} have been attributed to the vibration of OH groups v(O–H) of adsorbed water molecules [30], which disappeared after increasing the calcination temperature (as shown in Figure S2).

The FTIR spectrum of GCN shows numerous bands in the 1100–1630 cm^{-1} region corresponding to the typical stretching modes of GCN heterocycles. The FTIR spectra of the heterostructures NBCN-X(X = 1–4) exhibited characteristic peaks of NBO and GCN. However, for NBCN-1 and NBCN-2 the GCN band's peaks were found to be weaker than NBCN-3 and NBCN-4, due to the lower amount of GCN incorporated in the heterostructure. The results were consistent with the XRD results. Moreover, the sharp band at 810 cm^{-1} was associated with the tri-s-triazine forming the GCN structure. The broad band around 3163 cm^{-1} could be associated with the N–H or O–H bonds of the residual amino groups or absorbed H$_2$O molecules [31,32]. Detailed FTIR spectra are also provided in Table S1.

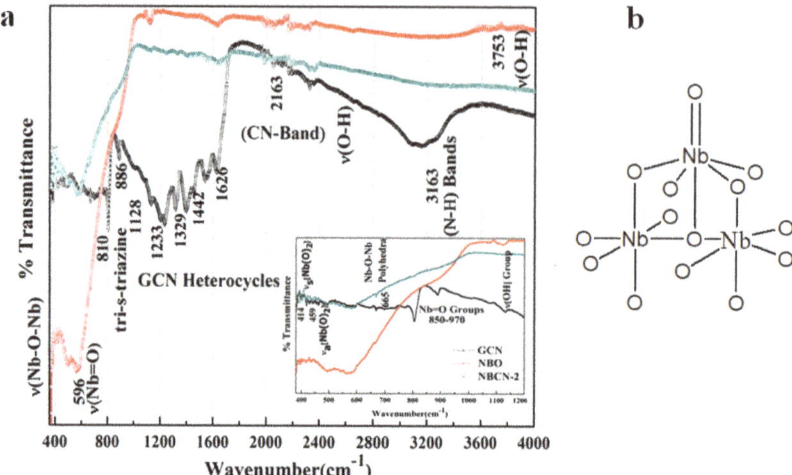

Figure 5. (**a**) FTIR spectra of NBO, GCN and NBCN-2 and (**b**) Highly and slightly distorted NbO_6 octahedra coexist in the structure.

SEM and TEM images of as-synthesized photocatalysts are shown in Figure 6. The bulk of the GCN and the hierarchical nanospheres of NBO can be seen in Figure 6a,b, respectively, while the NBCN heterostructures are given in Figure 6c–f. A significant change in the morphology can be observed for the NBO and NBCN heterostructures. BET analysis helped to further analyze the change in the surface area of bare and heterostructure photocatalysts. The high surface area of NBO-BA gradually decreased with increasing the annealing temperature from 200 to 500 °C (see Table S2). A high annealing temperature favored pore coalescence due to the crystallisation of walls separating mesopores in their structures. The BET surface areas of NBCN-X (X = 1–4) annealed at 200 °C and are provided in Table S3. The specific surface area decreased gradually with the increase in the amount of GCN, due to the low surface area of GCN.

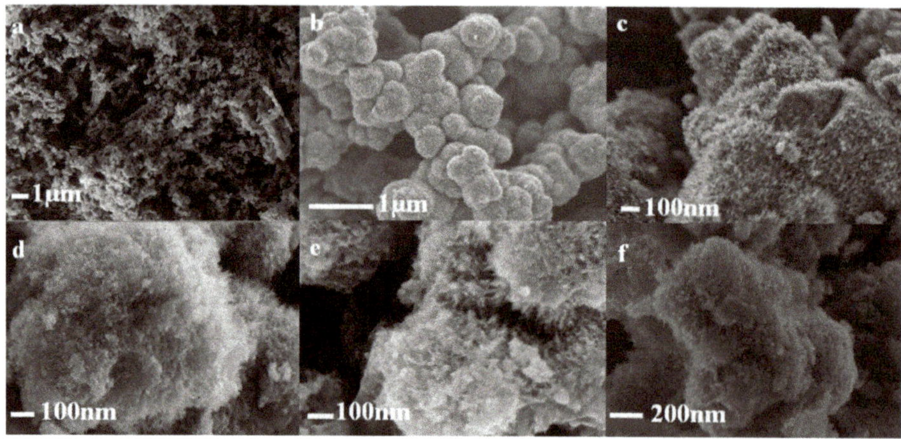

Figure 6. SEM images of (**a**) GCN, (**b**) NBO, (**c**) NBCN-1, (**d**) NBCN-2, (**e**) NBCN-3, and (**f**) NBCN-4.

A TEM analysis for NBO and NBCN-2 was conducted, as shown in Figure 7. For NBO (Figure 7a), the lattice fringes had 0.39 nm d-spacing's corresponding to the (001) lattice plane of Nb_2O_5. A small number of lattice fringes were observed due to the amorphous nature of NBO. The SAED pattern

(Figure 7a inset) indicated that the NBO was in the transforming state from the amorphous to the crystalline phase (consistent with XRD results). More TEM images at different resolutions are provided in Figure S3. For NBCN-2 (Figure 7b), lattice fringes for NBO and GCN were observed. However, due to a smaller amount of GCN, only a few lattice fringes for GCN were observed (0.31 nm d-spacing of (002) plane). The (001) plane of NBO was found to be compatible with the (002) plane of GCN, thus favoring the in-situ growth of NBO on the surface of GCN.

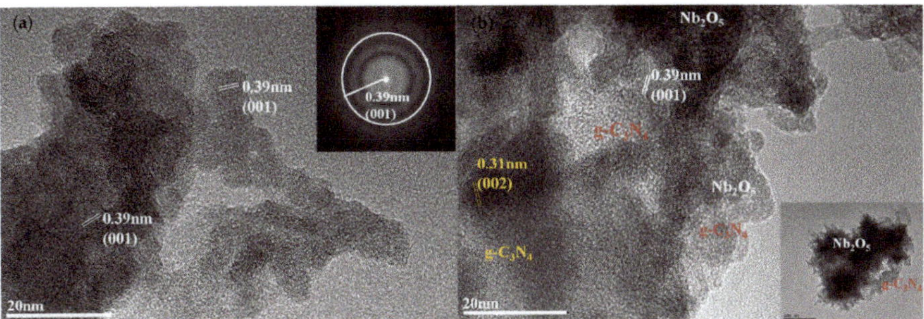

Figure 7. (a) HRTEM (high electron transmission electron microscopy) image and SAED (selected area electron diffraction) pattern (inset) of NBO, (b) HRTEM image and TEM (inset) image of NBCN-2.

2.3. Photocatalytic Evolution of Molecular H_2

A systematic study was conducted to signify the role of all possible parameters that could change the rate of generated molecular hydrogen (H_2). Initially, the photocatalytic formations of molecular H_2 for P25, GCN, NBO and NBCN-X (X = 1–4) were studied with and without Pt deposition in the presence of TEOA (a hole scavenger). The evolved H_2 is shown in Figure 8a,b, respectively. The total liberated amount in 7 h is also shown in Figure 8c. All heterostructures prepared generated more molecular H_2 than P25, NBO or GCN, both with and without Pt deposition. This increase in the H_2 formation rate with Pt deposition can be taken as evidence for the successful interfacial charge separation. Among all heterostructures, NBCN-2 showed the highest H_2 generation rate of 110 mmol/g·h, which could be associated with both its high surface area and suitable band positions. The evolution rates for other prepared composites: NBCN-1 = 103.27 mmol/g·h, NBCN-3 = 77.88 mmol/g·h, and NBCN-4 = 56.04 mmol/g·h, are also higher than those for P25 = 45.46 mmol/g·h, GCN=33.46 mmol/g·h or NBO= 41.20 mmol/g·h. Thus, the enhanced photocatalytic activity may have been caused by GCN loading acting as a visible light active material and as an efficient electron-hole separator for NBCN prepared heterostructures. However, this effect was only obvious provided that the optimised ratio of GCN and NBO was used.

For P25, the formation rate of molecular H_2 was increased in the first 4 h then a gradual decrease was observed. In comparison to bare materials the H_2 formation rate of P25 was larger than GCN and NBO, i.e., 45.46 mmol/g·h, but the rate was not stable after Pt deposition, as shown in Figure 8a,b, respectively. The H_2 formation rate was also smaller than all heterostructures prepared and the ~66% increase in molecular H_2 evolution rate was observed after Pt deposition, which was less than all heterostructures prepared. Moreover, the prepared composites showed more of a stable increase in the molecular H_2 formation rate than P25. The effect of methanol over platinized P25 could be studied through the Kandiel papers [33,34].

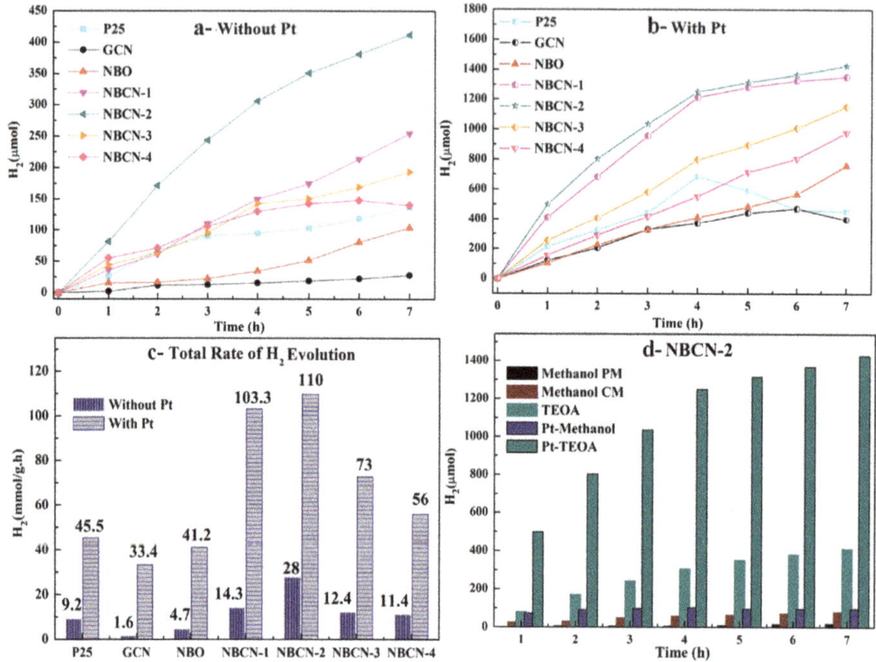

Figure 8. The photocatalytic formation of molecular H_2 of P25, NBO, GCN, NBCN-1, NBCN-2, NBCN-3 and NBCN-4: (**a**) without Pt, (**b**) with Pt deposition, and (**c**) total evolution rate; (**d**) photocatalytic formation of molecular H_2 of NBCN-2, in the presence of methanol and TEOA.

Secondly, to understand the role of the scavenger, the NBCN-2 activity employing methanol as hole scavenger was also investigated with and without Pt deposition. The change in H_2 evolution with time and in comparison with TEOA is shown in Figure 8d. A total of 5.53 mmol/g·h and 9.3 mmol/g·h of H_2 were liberated without and with Pt deposition, respectively. This increase was just 25%, and was 60% with TEOA. Moreover, in the presence of methanol, the H_2 amount was 67% (without Pt) and 85% (with Pt) less than TEOA. In light of the above results, TEOA not only gave higher activity, but also favored interfacial charge separation after Pt deposition. The findings indicated a possible change in the photocatalytic scheme. NBCN-2 results have also been compared with the physically mixed compound (with the same composition) in the presence of methanol. The NBCN-2 photocatalyst and physically mixed photocatalyst with the same ratio of NBO and GCN were used. A drastic change in the evolution rate was observed, as shown in Figure 8d, and more clearly in Figure S4. The NBCN-2 molecular H_2 generation rate showed a 35% increase over the physically mixed photocatalyst. Thus, in-situ heterostructures have higher activity than physically mixed heterostructures, and this can be attributed to good intimate contact. The total generated amount of H_2 for all studied photocatalysts has been provided in Table S4.

The present results have also been compared with previous reports on GCN, NBO and NBCN in the presence of TEOA/Methanol. The comparison is provided in Table 1. A detailed analysis of the table has proven the significance of our work in the following ways: (a) higher molecular H_2 rate in less time, (b) less concentration of photocatalysts and (c) the synthesized heterostructures have a higher molecular H_2 rate than many of them, even without the Pt deposition.

Table 1. A brief comparison of present molecular H_2 evolution rate employing g-C_3N_4 and/or Nb_2O_5.

Photocatalysts	Co-Catalyst	Scavenger	Reaction Conditions [a]	Light Source	H_2 (μmol/h/g)	Ref
Nb_2O_5 g-C_3N_4 P25 Nb_2O_5/g-C_3N_4 (Best Composite)	Pt (1.0 wt %)	TEOA (10 Vol. %)	45 mL, 10 mg (7 h)	Simulated Solar Lamp	41,200 33,460 45,460 110,000	Present
Cu_2O/g-C_3N_4-Mechanical Composite Cu_2O (0.05 wt %)-g-C_3N_4-In-Situ Composite g-C_3N_4	Pt (1.0 wt %)	TEOA (10 Vol. %)	180 mL, 100 mg (11 h)	Visible Light	142 241 142	[35]
Pt-g-C_3N_4-TiO_2 g-C_3N_4-Pt-TiO_2	Pt (not defined)	TEOA (10 Vol. %)	100 mL, 100 mg	Visible Light (420 nm)	1240 1780	[36]
Urea-polymerized-g-C_3N_4	Pt (3.0 wt %)	TEA (10 Vol. %) nil	100 mL, 80 mg (8 h) 100 mL, 80 mg (25 h)	Simulated Solar Lamp	590 3.13	[37]
Zn-tri-PcNc sensitized g-C_3N_4	Pt (1.0 wt %)	TEOA (10 Vol. %)	10 mL H_2O + 50 mL Ascorbic Acid, 10 mg (10 h)	Infrared 500 nm	12,500	[38]
g-C_3N_4 NiS/g-C_3N_4	Pt (2.0 wt %) NiS (1.5 mol %)	TEOA (10 Vol. %)	100 mL, 100 mg	Visible Light (420 nm)	480 450	[39]
g-C_3N_4	Pt (3.0 wt %)	TEOA (10 Vol. %) + K_2HPO_4	270 mL, 50 mg	Visible Light (420 nm)	18,940	[40]
Nb_2O_5/g-C_3N_4 (Best Composite) Nb_2O_5/g-C_3N_4 (Physically Mixed)	— Pt (1.0 wt %)	Methanol (10 Vol. %)	45 mL, 10 mg (7 h)	Simulated Solar Lamp	5530 9322 851	Present
Mesoporous Nb_2O_5	Pt (0.5 wt %)	Methanol (12.5 Vol. %)	400 mL, 10 mg (7 h)	Mercury Lamp (200–600 nm)	12,350	[41]
Mesoporous Nb_2O_5 nano particles	Pt (1.0 wt %)	Methanol (9.1 Vol. %)	220 mL, 20 mg (4 h)	Mercury Lamp Internal Irradiation	191	[42]
Mesoporous Nb_2O_5	— Pt (2.0 wt %) CuO NiO	Methanol (1M) + H_2SO_4 (1M)	200 mL, 200 mg (6 h)	White Light Source	1120 510 1405 800	[43]
Mesoporous Nb_2O_5	— Au (1.0 wt %) Pt (1.0 wt %) Cu (1.0 wt %) NiO (1.0 wt %) C	MeOH: H_2O = (1:5)	550 mL, 200 mg (6 h)	UV Light (360 nm) Internal Irradiation	328 2091 4647 1572 709 8.5	[44]

[a] Reaction Conditions: volume of photocatalytic solution, amount of photocatalyst and total time for H_2 evolution.

To estimate the electronic structure and possible interfacial band bending of materials, we conducted Mott–Schottky measurements (C^{-2} versus applied potential). The flat band/conduction band potentials (E_{FB}/E_{CB}), intercepted for NBO and GCN, are -0.61 V (-0.69) and -1.6 (-1.68) V vs. NHE (pH = 7), as shown in Figure 8a. The values have been converted to pH = 7 by using equations Equation (S1) and Equation (S2), respectively. By using the bandgap energy calculated through UV-vis-Absorption, the estimated valence band potentials (E_{VB}) are 2.32 V and 0.88 V for NBO and GCN at pH = 7, respectively (provided in Figure S5). Judging by the bandgap positions, typically a type-II mechanism should be followed, e.g., the photogenerated electrons in GCN could easily move to the NBO conduction band following the reduction there, and the photogenerated holes could easily move from the NBO valence band to the GCN valence band following the oxidation process. However, based on the recent report by Huang, Z.-F et al. [22], when TEOA is adsorbed on the surface the electron transfer is inverted as shown in Figure 1, and hence the Z-Scheme mechanism is followed. Following their results, the NBO electrons recombined with the holes of the GCN and consequently, reduction and oxidation reactions at the GCN and NBO occurred, respectively. The high production rate also favoured the direct Z-Scheme.

We estimated the charge carrier densities (N_D) to understand the interfacial band bending by using the Mott–Schottky slope (Figure 9). For a theoretical overview, the inclination angle of the M–S plot fo GCN was smaller than NBO. Since the relative dielectric constant (ε) was directly proportional to the inclination angle (θ) and inversely proportional to donor density (N_D) of the material. Hence, GCN should have a smaller dielectric constant and a higher donor density than NBO. In the literature, $\varepsilon \sim 7$–8 for GCN [32] and $\varepsilon \sim 38$ for pseudohexagonal NBO [45] has been reported. GCN ε is six times larger than NBO, so the N_D for GCN should be around six times larger than NBO. Mathematically, N_D has been calculated by using Equation (1) [46]:

$$N_D = \frac{2}{e_0 \varepsilon_0 \varepsilon} \left[\frac{d(1/C^2)}{dV} \right]^{-1}$$

(1)

where, e_0, ε_0, ε and $\frac{d(1/C^2)}{dV}$ are electron charge, vacuum permittivity, material dielectric constant and Mott–Schottky slope, respectively. The approximated values for GCN $N_D \approx 6.1 \times 10^{21}$ cm^{-3} and for NBO $N_D \approx 7.9 \times 10^{20}$ cm^{-3}, have been calculated, i.e., NBO N_D is seven times smaller than GCN, as expected.

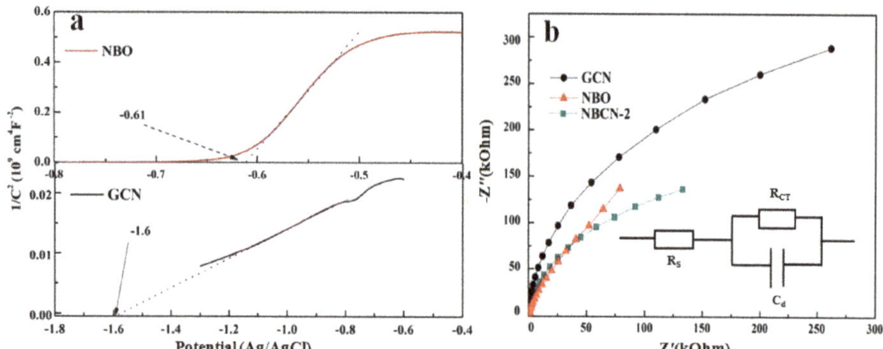

Figure 9. (a) Mott–Schottky plot for NBO and GCN at 1000 Hz vs. Ag/AgCl, (b) EIS analysis for NBO, GCN and NBCN-2.

For the n-type, due to high donor density, the Fermi level lay close to the bottom of the conduction band. Eventually, upward band bending occurred due to the high electron density of the conduction band and low acceptor density of the valence band. On the other hand, for low donor density of n-type, the Fermi level lay close to the middle of the bandgap. Moreover, a downward band bending of the conduction band and the valence band occurred. Thus, following the concept, after irradiation, the equilibrium Fermi level (E_F) of NBO and GCN was lying close to the redox potential (-0.41 V) vs. NHE (pH = 7). However, to justify the Z-Scheme and high production rate, fast recombination at the interface and fast charge transfer at the proposed conduction and valence bands should be followed.

Electrochemical Impedance Spectroscopy (EIS) analysis helped to further estimate the recombination and charge transfer process at the working electrode and electrolyte interfaces. Nyquist plots for NBO, GCN and NBCN-2 in the frequency range of 1000 kHz–0.01 Hz under UV-Vis light irradiation have been recorded (Figure 8b). The Randles circuit model [47] has been employed to describe the behavior of the electrode, as shown in the inset of Figure 8b. The arc radius has been associated with charge transfer resistance (R_{CT}) across the interface of working electrode/electrolyte, i.e., the small radius means efficient interfacial charge transfer/slow recombination rate of photogenerated electron/hole pairs. All three electrodes showed explicit arcs and the fitted values of R_{CT} are: NBO = 342 kΩ, GCN = 348 kΩ and NBCN-2 = 159 kΩ. The lowest R_{CT} for NBCN-2 has been recorded, which indicated that the composites have better efficiency of charge transfer than NBO and GCN. The NBO showed slightly better photoactivity than GCN. The large R_{CT} for GCN indicated its poor charge transfer characteristics, which may be associated with a low charge transfer rate leading to the fast recombination of photogenerated electron/hole pairs and poor photoactivity.

Thus, there was two times reduced resistance for the NBCN-2 heterostructures than NBO and GCN, which favored the reduced charge-transfer barrier with an increase in charge carrier density, and hence the Z-scheme system is followed. The proposed energy diagram and mechanism is shown in Figure 10. The proposed mechanism has been associated with the compatible band positions of NBO and GCN, which favored the direct Z-scheme transfer of charges and thus a higher molecular H_2 production rate. That is, the 1 wt % platinizied NBCN heterostructure exhibits a high activity for H_2 generation of 110 mmol/g·h because the direct transfer and recombination of photogenerated electrons in NBO and holes in GCN could greatly extend the lifetime time of carriers. The negative shift of the flat band of NBCN heterojunction (provided in Figure S6) further confirms the proposed Z-scheme mechanism.

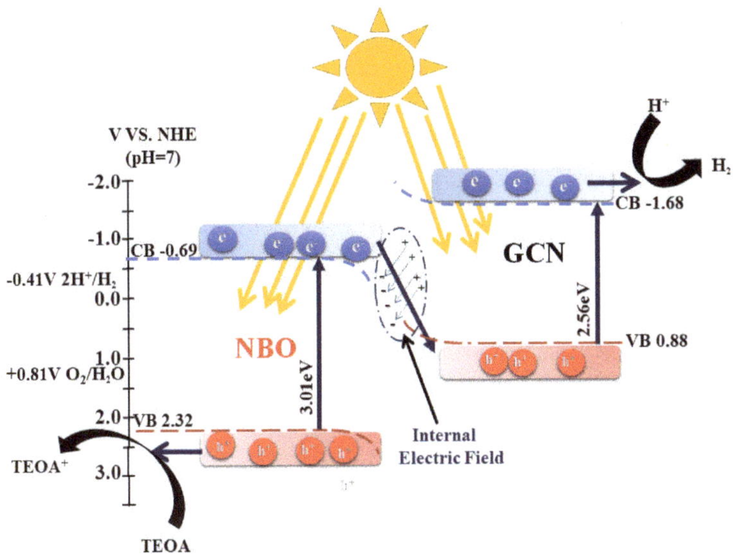

Figure 10. Schematic illustration of possible photocatalytic mechanism.

3. Methods

Niobium pentachloride $NbCl_5$ (99.9%, Sigma Aldrich, Munich, Germany), melamine $C_3H_6N_6$ (99.0%, Sigma Aldrich, Munich, Germany), hydrogen peroxide solution H_2O_2 (30 wt %), nitric acid HNO_3 (60 wt %), Evonil Aeroxide TiO_2 P25, chloroplatinic acid hexahydrate $H_2PtCl_6.6H_2O$ ($\geq$37.05%, Sigma Aldrich, Munich, Germany) and triethanolamine $(HOCH_2CH_2)_3N$ (99.5%, Sigma Aldrich, Munich, Germany) were used without further purification.

3.1. Material Synthesis

3.1.1. Preparation of Nb_2O_5 (NBO)

0.5 g of hygroscopic yellow powder of $NbCl_5$ was added into a mixture of 20 mL de-ionised water and 0.5 mL HNO_3. The solid dissolved immediately yielding a transparent solution. 10 mL of aqueous H_2O_2 solution was added drop-wise into a clear solution under vigorous stirring. The solution became light yellow, confirming the formation of the niobium-peroxo complex (NPC). This solution was diluted to 60 mL by further addition of de-ionised water. The prepared light-yellow solution was subesquently transferred into a 250 mL autoclave and maintained at 110 °C for 24 h in an oven. The synthesised white precipitate was centrifuged and washed several times with de-ionised water. Subsequently, the prepared powder was dried in an oven at 90 °C for 12 h (named as NBO-BA). The resultant NBO-BA samples were further annealed in a muffle furnace at different temperatures ranging from 200 °C to 500 °C for 2 h with a heating rate of 10 °C min^{-1}. In the present work, the photocatalyst annealed at 200 °C has been named as NBO, throughout.

3.1.2. Preparation of g-C_3N_4 (GCN)

10 g of melamine were placed in a crucible with a cover lid on top and then annealed at 550 °C for 3 h in a muffle furnace employing a heating rate of 10 °C min^{-1}. After natural cooling, a yellow powder of bulk g-C_3N_4 was obtained (named as GCN).

3.1.3. Preparation of Nb_2O_5/g-C_3N_4 (NBCN)

The Nb_2O_5/g-C_3N_4 heterostructures named as NBCN-X (where X = 1, 2, 3, 4) were prepared by using different amounts of GCN. Typically, 2.5 mg of NBO-BA were obtained by hydrothermal synthesis. Therefore, for the heterostructure preparations, the selected amount of GCN was added to the prepared solution of NBO-BA, i.e., X: 1 = 5 wt %, 2 = 10 wt %, 3 = 30 wt % and 4 = 60 wt %. The synthesis process was as follows: initially, GCN suspension was prepared in 20 mL of de-ionised water following the ultra-sonication for 30 min. Then, the prepared GCN suspension was added to a light-yellow solution of NPC under continuous stirring. The amount was adjusted to 60 mL by further addition of de-ionised water and stirring for 30 min. Choosing pH = 0.7 should assure the positive and negative surface charges have been developed over GCN and NPC, respectively, for favorable interface development. The prepared concentrated solution was subsequently transferred to a 250 mL autoclave and maintained at the 110 °C for 24 h in the oven. After washing and drying at 90 °C for 12 h, a light-yellow powder is obtained, which was further annealed at 200 °C for 2 h in a muffle furnace with a 10 °C min^{-1} ramp rate.

3.2. Photodeposition of Platinum (Pt)

The photo-deposition technique was applied to deposit 1 wt % platinum (Pt) on the samples. Hexachloroplatinic acid ($H_2PtCl_6 \cdot 6H_2O$) was used as the Pt precursor and methanol as a reducing agent. The calculated amounts of $H_2PtCl_6 \cdot 6H_2O$ and the photocatalyst were added to 10 vol/vol. % of aqueous methanol solution. The resulting suspension was transferred into a closed reactor and placed under UV-light for 12 h with continuous stirring. The suspension was washed several times with de-ionised water to remove non-deposited Pt and subsequently with ethanol. The obtained Pt loaded photocatalysts were dried in an oven for 12 h at 90 °C.

3.3. Material Characterization

A Bruker (D8 Advance) instrument using Cu Kα (α = 0.15406 nm) radiation was used to record the X-ray diffraction (XRD) data of the as-synthesized samples. Scanning electron microscopy (SEM, JEOL JSM-6700F, JEOL, Tokyo, Japan) with a LEI detector (Lower Secondary Electron Image) was employed to analyze the morphologies. Transmission electron microscopy (TEM), using an FEI Tecnai G2 F20 microscope operating at 200 kV was used to characterise the samples further. Micrographs were taken in bright field (BF) and in a selected area electron diffraction mode. A Varian Cary 100 Bio was used to measure the UV–vis absorption spectra. A Bruker Vertex 80v spectrophotometer (Bruker, Billerica, MA, USA) was used to measure the FTIR spectra from 4000 to 400 cm^{-1} in vacuum.

3.4. Photocatalytic Molecular Hydrogen (H$_2$) Formation

The prepared photocatalysts without and with Pt deposits were used to conduct photocatalytic molecular hydrogen (H$_2$) evolution reactions. In a typical experimental run, 0.01 g of the photocatalyst were added into 45 mL of de-ionised water (maintaining pH 5.6). Methanol or TEOA were added resulting in a suspension which contains 10 vol % of a hole scavenger. The suspension inside the photoreactor was thoroughly degassed for 30 min with Ar to remove air and then stirred for 30 min in the dark to establish the adsorption equilibrium. Afterwards, the photoreactor was placed under a Xenon light source (1000 W, 1.5G) suitable to simulate solar light for 7 h. The temperature was maintained constant by using a homemade cooling system. The evolved amounts of H$_2$ were measured every 1 h by using a gas chromatograph (Shimadzu 8A (Shimadzu, Kyoto, Japan) equipped with a TCD detector and a 5 Å molecular sieve packed column; Ar was used as the carrier gas).

3.5. Photoelectrochemical Measurements

The photoelectrochemical (PEC) measurements were conducted by using an electrochemical workstation (CHI-660B, CH Instruments, Inc., Austin, TX, USA) accompanying a ZAHNER PECC-2 reactor and 450W xenon lamp as a light source. Moreover, a standard three-compartment cell (consisting of a photo-/working electrode (WE), a Pt wire counter electrode (CE) and an Ag/AgCl reference electrode (RE)) with 0.2 M Na$_2$SO$_4$ electrolyte solution (pH = 5.6) were used. The working electrode was prepared using a screen-printing method and then annealed at 400 °C for 2 h to remove organic chemicals. Mott–Schottky measurements were performed at a frequency range of 10–1000 Hz with 10 mV amplitude. Electrochemical impedance spectra were obtained under irradiation at open circuit voltage over a frequency range from 1000 Hz to 0.01 Hz, with an AC voltage at 250 mV vs. Ag/AgCl reference electrode.

4. Conclusions

Z-scheme Nb$_2$O$_5$/g-C$_3$N$_4$ heterostructures with excellent molecular H$_2$ production activity were prepared via in-situ hydrothermal syntheses. The prepared heterostructures exhibited excellent photocatalytic activity compared to individual g-C$_3$N$_4$ and Nb$_2$O$_5$ under simulated solar light illumination. The highest reported H$_2$ evolution rate was 110 mmol/g·h (7.7 mmol). We found that by increasing the amount of g-C$_3$N$_4$, the molecular H$_2$ production rate decreased, indicating more intimidating interface development does not favor photocatalytic reactions. However, the molecular H$_2$ evolution rate for all prepared heterostructures was higher than many other semiconductors reported in the literature. Moreover, we justified our results by a reduced recombination rate, high charge carrier density and complemented band positions. For future work, we suggest that the various heterojunction materials possessing diverse structural morphology exhibiting a higher photocatalytic activity be prepared by the simple methodology described here. These studies will pave the way for a new dimension in photocatalytic studies of Nb_2O_5 and g-C_3N_4 nanocomposites for enhanced molecular H_2 production.

Supplementary Materials: The following are available online at http://www.mdpi.com/2073-4344/9/2/169/s1, Figure S1: XRD patterns BA and AA at different calcination temperatures, Figure S2: FTIR spectra of all photocatalysts, Figure S3: TEM images of NBO, Figure S4: Molecular H_2 generation of NBCN-2, Chemical and Physical Mixing in the presence of methanol, Figure S5: Bandgap vs. photon energy by UV-vis-Diffuse Absorption Spectra of a) NBO, GCN, NBCN-1, NBCN-2, NBCN-3, and NBCN-4 and b) with the change in the annealing temperature, Figure S6: Mott-Schottky plot of NBO and NBCN-2, Table S1: Associated Bands in FTIR Spectra, Table S2: Specific surface area (m^2/g) of samples with different calcination temperatures, Table S3: Specific surface area of NBO and NBCN-X (X = 1–4) without calcination, Table S4: Liberated Amount of H_2 after 7h with and without Pt, Equation (S1) and Equation (S2).

Author Contributions: F.I. conceived, designed and performed the experiments; F.K.B. and M.T. analyzed the data; R.D. contributed reagents/materials/analysis tools; F.I., R.D. and D.B. wrote the paper. Substantial paper changes have been made by Ralf and Bahnemann.

Funding: The work was supported by the Alexendar Von Humboldt Foundation (Project No. 60421802) and PSF/NSFC/Eng-P-UoL(02).

Conflicts of Interest: We declare no conflict of interest.

References

1. Fujishima, A.; Honda, K. Electrochemical Photolysis of Water at a Semiconductor Electrode. *Nature* **1972**, *238*, 37. [CrossRef] [PubMed]

2. Yu, J.; Qi, L.; Jaronie, M. Hydrogen Production by Photocatalytic Water Splitting over Pt/TiO_2 Nanosheets with Exposed (001) Facets. *J. Phys. Chem. C* **2010**, *114*, 13118–13125. [CrossRef]

3. Cho, I.S.; Chen, Z.; Forman, A.J.; Kim, D.R.; Rao, P.M.; Jaramillo, T.F.; Zheng, X. Branched TiO_2 Nanorods for Photoelectrochemical Hydrogen Production. *Nano Lett.* **2011**, *11*, 4978–4984. [CrossRef] [PubMed]

4. Luo, J.; Steier, L.; Son, M.K.; Schreier, M.; Mayer, M.T.; Michael, G. Cu_2O Nanowire Photocathodes for Efficient and Durable Solar Water Splitting. *Nano Lett.* **2016**, *16*, 1848–1857. [CrossRef] [PubMed]

5. Ye, R.; Fang, H.; Zheng, Y.-Z.; Li, N.; Wang, Y.; Tao, X. Fabrication of $CoTiO_3/g$-C_3N_4 Hybrid Photocatalysts with Enhanced H_2 Evolution: Z-Scheme Photocatalytic Mechanism Insight. *ACS Appl. Mater. Interfaces* **2016**, *8*, 13879–13889. [CrossRef] [PubMed]

6. Tahir, M.; Cao, C.; Butt, F.K.; Butt, S.; Idrees, F.; Ali, Z.; Aslam, I.; Tanveer, M.; Mahmood, A.; Mahmood, N. Large scale production of novel g-C_3N_4 micro strings with high surface area and versatile photodegradation ability. *CrystEngComm* **2014**, *16*, 1825–1830. [CrossRef]

7. Tahir, M.; Cao, C.; Butt, F.K.; Idrees, F.; Mahmood, N.; Ali, Z.; Aslam, I.; Tanveer, M.; Rizwan, M.; Mahmood, T. Tubular graphitic-C_3N_4: A prospective material for energy storage and green photocatalysis. *J. Mater. Chem. A* **2013**, *1*, 13949–13955. [CrossRef]

8. Yan, J.; Wu, G.; Guan, N.; Li, L. Nb_2O_5/TiO_2 heterojunctions: synthesis strategy and photocatalytic activity. *Appl. Catal. B* **2014**, *152*, 280–288. [CrossRef]

9. Tahir, M.; Cao, C.; Mahmood, N.; Butt, F.K.; Mahmood, A.; Idrees, F.; Hussain, S.; Tanveer, M.; Ali, Z.; Aslam, I. Multifunctional g-C_3N_4 nanofibers: A template-free fabrication and enhanced optical, electrochemical, and photocatalyst properties. *ACS Appl. Mater. Interfaces* **2013**, *6*, 1258–1265. [CrossRef]

10. Sridharan, K.; Jang, E.; Park, T.J. Novel visible light active graphitic C_3N_4–TiO_2 composite photocatalyst: Synergistic synthesis, growth and photocatalytic treatment of hazardous pollutants. *Appl. Catal. B* **2013**, *142–143*, 718–728. [CrossRef]

11. Hou, Y.; Zhu, Y.; Xu, Y.; Wang, X. Photocatalytic hydrogen production over carbon nitride loaded with WS_2 as cocatalyst under visible light. *Appl. Catal. B* **2014**, *156–157*, 122–127. [CrossRef]

12. Bai, Y.; Wang, P.-Q.; Liu, J.-Y.; Liu, X.-J. Enhanced photocatalytic performance of direct Z-scheme $BiOCl$-g-C_3N_4 photocatalysts. *RSC Adv.* **2014**, *4*, 19456–19461. [CrossRef]

13. Aslam, I.; Cao, C.; Tanveer, M.; Khan, W.S.; Tahir, M.; Abid, M.; Idrees, F.; Butt, F.K.; Ali, Z.; Mahmood, N. The synergistic effect between WO_3 and g-C_3N_4 towards efficient visible-light-driven photocatalytic performance. *New J. Chem.* **2014**, *38*, 5462–5469. [CrossRef]

14. Katsumata, H.; Tachi, Y.; Suzuki, T.; Kaneco, S. Z-scheme photocatalytic hydrogen production over WO_3/g-C_3N_4 composite photocatalysts. *RSC Adv.* **2014**, *4*, 21405–21409. [CrossRef]

15. Acar, C.; Dincer, I.; Naterer, G.F. Review of photocatalytic water-splitting methods for sustainable hydrogen production. *Int. J. Energy Res.* **2016**, *40*, 1449–1473. [CrossRef]
16. Idrees, F.; Cao, C.; Ahmed, R.; Butt, F.K.; Butt, S.; Tahir, M.; Tanveer, M.; Aslam, I.; Ali, Z. Novel nano-flowers of Nb_2O_5 by template free synthesis and enhanced photocatalytic response under visible light. *Sci. Adv. Mater.* **2015**, *7*, 1298–1303. [CrossRef]
17. Idrees, F.; Hou, J.; Cao, C.; Butt, F.K.; Shakir, I.; Tahir, M.; Idrees, F. Template-free synthesis of highly ordered 3D-hollow hierarchical Nb_2O_5 superstructures as an asymmetric supercapacitor by using inorganic electrolyte. *Electrochim. Acta* **2016**, *216*, 332–338. [CrossRef]
18. Lam, S.-M.; Sin, J.-C.; Satoshi, I.; Abdullah, A.Z.; Mohamed, A.R. Enhanced sunlight photocatalytic performance over Nb_2O_5/ZnO nanorod composites and the mechanism study. *Appl. Catal. A* **2014**, *471*, 126–135. [CrossRef]
19. Hong, Y.; Li, C.; Zhang, G.; Meng, Y.; Yin, B.; Zhao, Y.; Shi, W. Efficient and stable Nb_2O_5 modified g-C_3N_4 photocatalyst for removal of antibiotic pollutant. *Chem. Eng. J.* **2016**, *299*, 74–84. [CrossRef]
20. Huang, Q.-Z.; Wang, J.-C.; Wang, P.-P.; Yao, H.-C.; Li, Z.-J. In-situ growth of mesoporous Nb_2O_5 microspheres on g-C_3N_4 nanosheets for enhanced photocatalytic H_2 evolution under visible light irradiation. *Int. J. Hydrogen Energy* **2017**, *42*, 6683–6694. [CrossRef]
21. Low, J.; Jiang, C.; Cheng, B.; Wageh, S.; Al-Ghamdi, A.A.; Yu, J. A Review of Direct Z-Scheme Photocatalysts. *Small Methods* **2017**, *1*, 1700080. [CrossRef]
22. Huang, Z.-F.; Song, J.; Wang, X.; Pan, L.; Li, K.; Zhang, X.; Wang, L.; Zou, J.-J. Switching charge transfer of C_3N_4/$W_{18}O_{49}$ from type-II to Z-scheme by interfacial band bending for highly efficient photocatalytic hydrogen evolution. *Nano Energy* **2017**, *40*, 308–316. [CrossRef]
23. Huang, Y.-T.; Cheng, R.; Zhai, P.; Lee, H.; Chang, Y.-H.; Feng, S.-P. Solution-Based Synthesis of Ultrasmall Nb_2O_5 Nanoparticles for Functional Thin Films in Dye-Sensitized and Perovskite Solar Cells. *Electrochim. Acta* **2017**, *236*, 131–139. [CrossRef]
24. Passoni, L.C.; Siddiqui, M.R.H.; Steiner, A.; Kozhevnikov, I.V. Niobium peroxo compounds as catalysts for liquid-phase oxidation with hydrogen peroxide. *J. Mol. Catal. A: Chem.* **2000**, *153*, 103–108. [CrossRef]
25. Narendar, Y.; Messing, G.L. Synthesis, decomposition and crystallization characteristics of peroxo− citrato−niobium: an aqueous niobium precursor. *Chem. Mater.* **1997**, *9*, 580–587. [CrossRef]
26. Silva, G.; Carvalho, K.; Lopes, O.; Ribeiro, C. g-C3N4/Nb2O5 heterostructures tailored by sonochemical synthesis: Enhanced photocatalytic performance in oxidation of emerging pollutants driven by visible radiation. *Appl. Catal. B Environ.* **2017**, *216*, 70–79.
27. Lopes, O.; Paris, E.; Ribeiro, C. Synthesis of Nb_2O_5 Nanoparticles Through the Oxidant Peroxide Method Applied to Organic Pollutant Photodegradation: A Mechanistic Study. *Appl. Catal. B* **2014**, *144*, 800–808. [CrossRef]
28. Jehng, J.M.; Wachs, I.E. Niobium oxide solution chemistry. *J. Raman Spectrosc.* **1991**, *22*, 83–89. [CrossRef]
29. Compton, O.C.; Osterloh, F.E. Niobate nanosheets as catalysts for photochemical water splitting into hydrogen and hydrogen peroxide. *J. Phys. Chem. C* **2008**, *113*, 479–485. [CrossRef]
30. Nakajima, K.; Fukui, T.; Kato, H.; Kitano, M.; Kondo, J.N.; Hayashi, S.; Hara, M. Structure and Acid Catalysis of Mesoporous $Nb_2O_5\cdot nH_2O$. *Chem. Mater.* **2010**, *22*, 3332–3339. [CrossRef]
31. Cui, Y.; Ding, Z.; Liu, P.; Antonietti, M.; Fu, X.; Wang, X. Metal-free activation of H_2O_2 by g-C_3N_4 under visible light irradiation for the degradation of organic pollutants. *Phys. Chem. Chem. Phys.* **2012**, *14*, 1455–1462. [CrossRef] [PubMed]
32. Giannakopoulou, T.; Papailias, I.; Todorova, N.; Boukos, N.; Liu, Y.; Yu, J.; Trapalis, C. Tailoring the energy band gap and edges' potentials of g-C_3N_4/TiO_2 composite photocatalysts for NO_x removal. *Chem. Eng. J.* **2017**, *310*, 571–580. [CrossRef]
33. Kandiel, T.A.; Dillert, R.; Robben, L.; Bahnemann, D.W. Photonic efficiency and mechanism of photocatalytic molecular hydrogen production over platinized titanium dioxide from aqueous methanol solutions. *Catal. Today* **2011**, *161*, 196–201. [CrossRef]
34. Kandiel, T.A.; Ivanova, I.; Bahnemann, D.W. Long-term investigation of the photocatalytic hydrogen production on platinized TiO_2: An isotopic study. *Energy Environ. Sci.* **2014**, *7*, 1420–1425. [CrossRef]
35. Chen, J.; Shen, S.; Guo, P.; Wang, M.; Wu, P.; Wang, X.; Guo, L. In-situ reduction synthesis of nano-sized Cu_2O particles modifying g-C_3N_4 for enhanced photocatalytic hydrogen production. *Appl. Catal. B* **2014**, *152–153*, 335–341. [CrossRef]

36. Chai, B.; Peng, T.; Mao, J.; Li, K.; Zan, L. Graphitic carbon nitride (g-C_3N_4)-Pt-TiO_2 nanocomposite as an efficient photocatalyst for hydrogen production under visible light irradiation. *Phys. Chem. Chem. Phys.* **2012**, *14*, 16745–16752. [CrossRef] [PubMed]

37. Zhang, Y.; Liu, J.; Wu, G.; Chen, W. Porous graphitic carbon nitride synthesized via direct polymerization of urea for efficient sunlight-driven photocatalytic hydrogen production. *Nanoscale* **2012**, *4*, 5300–5303. [CrossRef]

38. Zhang, X.; Yu, L.; Zhuang, C.; Peng, T.; Li, R.; Li, X. Highly Asymmetric Phthalocyanine as a Sensitizer of Graphitic Carbon Nitride for Extremely Efficient Photocatalytic H_2 Production under Near-Infrared Light. *ACS Catal.* **2014**, *4*, 162–170. [CrossRef]

39. Chen, Z.; Sun, P.; Fan, B.; Zhang, Z.; Fang, X. In Situ Template-Free Ion-Exchange Process to Prepare Visible-Light-Active g-C_3N_4/NiS Hybrid Photocatalysts with Enhanced Hydrogen Evolution Activity. *J. Phys. Chem. C* **2014**, *118*, 7801–7807. [CrossRef]

40. Liu, G.; Wang, T.; Zhang, H.; Meng, X.; Hao, D.; Chang, K.; Li, P.; Kako, T.; Ye, J. Nature-Inspired Environmental "Phosphorylation" Boosts Photocatalytic H_2 Production over Carbon Nitride Nanosheets under Visible-Light Irradiation. *Angew. Chem. Int. Ed.* **2015**, *54*, 13561–13565. [CrossRef]

41. Chen, X.; Yu, T.; Fan, X.; Zhang, H.; Li, Z.; Ye, J.; Zou, Z. Enhanced activity of mesoporous Nb_2O_5 for photocatalytic hydrogen production. *Appl. Surf. Sci.* **2007**, *253*, 8500–8506. [CrossRef]

42. Sreethawong, T.; Ngamsinlapasathian, S.; Lim, S.H.; Yoshikawa, S. Investigation of thermal treatment effect on physicochemical and photocatalytic H_2 production properties of mesoporous-assembled Nb_2O_5 nanoparticles synthesized via a surfactant-modified sol–gel method. *Chem. Eng. J.* **2013**, *215–216*, 322–330. [CrossRef]

43. Pai, Y.-H.; Fang, S.-Y. Preparation and characterization of porous Nb_2O_5 photocatalysts with CuO, NiO and Pt cocatalyst for hydrogen production by light-induced water splitting. *J. Power Sources* **2013**, *230*, 321–326. [CrossRef]

44. Lin, H.-Y.; Yang, H.-C.; Wang, W.-L. Synthesis of mesoporous Nb_2O_5 photocatalysts with Pt, Au, Cu and NiO cocatalyst for water splitting. *Catal. Today* **2011**, *174*, 106–113. [CrossRef]

45. Manuspiya, H. Electrical Properties of Niobium Based Oxides-Ceramics and Single Crystal Fibers Grown by the Laser-Heated Pedestal Growth (LHPG) Technique. Ph.D. Thesis, Penn State University, University Park, PA, USA, 4 April 2003.

46. Gelderman, K.; Lee, L.; Donne, S.W. Flat-Band Potential of a Semiconductor: Using the Mott–Schottky Equation. *J. Chem. Educ.* **2007**, *84*, 685. [CrossRef]

47. Mousa, G.; Golnaraghi, F.; DeVaal, J.; Young, A. Detecting proton exchange membrane fuel cell hydrogen leak using electrochemical impedance spectroscopy method. *J. Power Sources* **2014**, *246*, 110–116. [CrossRef]

Article

Synthesis of Spherical TiO₂ Particles with Disordered Rutile Surface for Photocatalytic Hydrogen Production

Na Yeon Kim, Hyeon Kyeong Lee, Jong Tae Moon and Ji Bong Joo *

School of Chemical Engineering, Konkuk University, Gwangjin-gu, Seoul 05029, Korea;
kny960403@konkuk.ac.kr (N.Y.K.); hyeonk@konkuk.ac.kr (H.K.L.); aidf91@gmail.com (J.T.M.)
* Correspondence: jbjoo@konkuk.ac.kr; Tel.: +82-245-03-545

Received: 28 April 2019; Accepted: 24 May 2019; Published: 28 May 2019

Abstract: One of the most important issues in photocatalysis research has been the development of TiO₂-based photocatalysts that work efficiently under visible light conditions. Here, we report the monodispersed, spherical TiO₂ particles with disordered rutile surface for use as visible-light photocatalysts. The spherical TiO₂ particles with disordered surface were synthesized by sol-gel synthesis, followed by sequential calcination, and chemical reduction process using Li/Ethylenediamine (Li/EDA) solution. Variation of the calcination temperature allowed the crystalline properties of the calcined TiO₂ samples, such as the ratio of anatase and rutile, to be finely controlled. The content ratios of anatase phase to rutile phase leads to different degrees of disorder of the rutile surface, which is closely related to the photocatalysis activity. Chemical reduction using the Li/EDA solution enables selective reduction of the rutile surface of the calcined TiO₂, resulting in enhanced light absorption. As a result, we were able to synthesize spherical TiO₂ photocatalysts having a disordered rutile surface in a mixed crystalline phase, which is beneficial during photocatalysis in terms of light absorption and charge separation. When used as photocatalysts for hydrogen production under solar light conditions, the chemically-reduced TiO₂ particles with both the disordered rutile surface and mixed crystalline phase showed significantly enhanced catalytic activity.

Keywords: TiO₂; spherical particle; disordered surface; photocatalysts; hydrogen production

---

1. Introduction

In modern society the energy crisis is becoming one of the biggest issues to directly impact our lives. Hydrogen as a green energy carrier has attracted much attention, due to its high energy capacity, environment-friendly characteristics, and sustainability [1]. As people are recognizing that high concentration of carbon dioxide is closely related to global warming and climate change, hydrogen can be considered as a one of the representative alternative-energy resources to either reduce or replace the use of depletable fossil fuels. There are several strategies to produce hydrogen, such as the reforming of either fossil fuel or renewable biomass, water electrolysis, ammonia decomposition, photo-electrochemical water splitting, and photocatalytic water splitting [2–8]. Among them, photocatalytic hydrogen production from water is considered as an ideal and economically-feasible method, since infinite solar energy can be used with any other type of energy resource [9].

Titanium dioxide (Titania, TiO₂) is one of the most well-known semiconductor photocatalysts. TiO₂ materials have a few advantages, which include low cost, considerable photocatalytic activity, low toxicity, high chemical stability, and abundance on Earth [10–12]. Since Honda and Fujishima first discovered hydrogen production by the photoelectrochemical splitting of water under UV light conditions [13], TiO₂ has not only been intensively studied with fundamental researches, but also widely used for practical systems for solar energy conversion [14–17]. Although TiO₂ has been intensively studied over the past decades and has become known as a superb photocatalyst, the practical use of TiO_2 has still been limited, due to its intrinsic optical property. Since TiO_2 has a wide band-gap energy of 3.0–3.2 eV, its photocatalytic performance is limited to the ultra-violet (UV) region [12,17,18]. Even though the greater portion of solar light is in the visible and IR region, TiO_2 can absorb mainly UV light, resulting in it generally showing low solar-to-chemical efficiency under solar light conditions. In addition, rapid recombination of photogenerated electron-hole pairs also leads to low quantum efficiency. Thus, this results in low overall photocatalytic activity [19,20].

In order to overcome the above drawbacks, various novel approaches have been taken to improve its optical, electronic, and chemical properties. To narrow the band-gap, either nonmetal or metal ions are doped into TiO_2 crystal lattice, regulating either the level of conduction band or valence bands [21–27]. Surface sensitization using organic dyes can allow TiO_2 to utilize the exited electron from dye molecules under visible light irradiation [28,29]. Recently, decoration of plasmonic metal nanoparticles on the TiO_2 surface have also been suggested to improve hot electron transfer from metal nanoparticles to the conduction band of TiO_2, resulting in unexpected photocatalytic performance under visible light conditions [30,31]. Since Mao and co-worker made the pioneering discovery of black TiO_2 nanocrystal through surface-disordering using hydrogen [32], there have been many further reports about the interesting strategy of the synthesis of surface-disordered TiO_2 by chemical reduction [33–35]. Unlike other TiO_2-based photocatalysts by conventional disordering processes, they can achieve significantly disordered surface of TiO_2 nanocrystal with well-maintained crystalline property, resulting in high photocatalytic activity on both organic dye decomposition and hydrogen production under solar light [32]. It is also well known that active metal, such as Al and Mg, can reduce TiO_2, resulting in the formation of colored TiO_2. Huang et al. developed a new approach to prepare colored TiO_2 based on Al-reduction [36]. Sinhamahapatra and Yu synthesized black TiO_2 by mixing the commercial TiO_2 nanocrystals with Mg powder, followed by annealing in the hydrogen environment [37,38]. Park et al. also developed chemical reduction using a Lithium/Ethylenediamine mixture, and prepared blue-colored TiO_2 that had the selectively disordered rutile surface. Lithium/Ethylenediamine (Li/EDA) solution, a metal chelate compound, is a very strong reducing agent. Ethylenediamine (EDA) is a superbase chemical that provides high pH conditions. The chelated metallic Li derived from the Li/EDA solution then selectively breaks the bonds between Ti and O of the rutile phase TiO_2, resulting in various defects and a disordered surface. Defects in the disordered surface can narrow the band gap, forming interstates between the conduction band and valence band. Therefore, white TiO_2 turns into a blueish color [39].

As shown in the previous literature, several chemical reduction approaches can allow the band-gap of pure TiO_2 to be narrow, and the colored TiO_2 to have high performance under solar light photocatalysis. Although intensive investigation of the synthesis of colored TiO_2 was conducted using commercial TiO_2 nanocrystals, such as P25 [39–41], there has been only limited study reported on the fabrication of colloidal TiO_2 nanostructure. We recently found that uniform colloidal TiO_2 particles with tunable crystalline properties, such as the ratio of anatase–rutile phase and crystallinity, can be produced by the sol-gel synthesis of titanium-alkoxide precursors, followed by calcination at different temperatures [42]. As-calcined TiO_2 samples show monodispersed colloid particles and well-developed TiO_2 crystallinity, with finely-tunable anatase–rutile mixed phase. As previously mentioned, Li/EDA solution can selectively reduce the rutile phase of TiO_2, and disorder its surface [39]. Since the crystalline phase of spherical TiO_2 particles between anatase and rutile can be finely tuned by varying the calcination temperature while maintaining the morphological dimension, the degree of surface disordering can be systemically controlled, resulting in precise control of the band-gap energy.

In this work, we report the synthesis of the spherical TiO_2 particle with the disordered rutile surface for photocatalytic hydrogen production. Specifically, monodispersed TiO_2 particles with tunable crystalline property are synthesized by sol-gel synthesis, followed by calcination at different temperatures. Since TiO_2 samples have different ratios of anatase to rutile phases, the optical properties are conveniently controlled by Li/EDA treatment. The resulting Li/EDA-treated TiO_2 samples showed advantageous characteristics, such as uniform particle dimension, favorable dispersity, facile absorption of visible light, and controllable degree of disorder of the surface. Corresponding with such reduced TiO_2 spherical photocatalysts, it was possible to achieve enhanced performance in photocatalytic hydrogen production under solar light irradiation. We systemically study and discuss the optical properties, physicochemical characteristics, and photocatalytic performance of the spherical TiO_2 with disordered rutile surface.

2. Results and Discussion

Colored TiO_2 spherical particles with the disordered rutile surface were synthesized by a modified sol-gel synthesis, followed by sequential calcination at the desired temperature and chemical reduction, respectively (Figure 1a). More specifically, the synthesis consisted of the following steps: (i) Preparation of a colloidal amorphous TiO_2 sphere (AT, amorphous TiO_2) (ii) calcination of a TiO_2 sphere to convert to the crystalline counterpart (CT-x); and (iii) chemical reduction of crystalline TiO_2, to the colored one (RT-X), by using Li/EDA (Li in ethylenediamine) as reducing agent. During preparation of the colloidal TiO_2 particle, the monodispersed amorphous TiO_2 spheres were synthesized by the sol-gel reaction of titanium n-butoxide (TBOT) in mixed solvent of ethanol and acetonitrile, in the presence of base ammonia and surfactant. The hydrolysis and condensation of the TBOT were highly influenced by several synthetic parameters, such as the solvent environment, the amount of precursor, and the concentration of surfactant. Recently, we systemically studied the effect of the synthetic parameters on the physical–chemical properties of colloidal TiO_2 particles, and successfully synthesized uniform TiO_2 spheres with controllable crystalline properties [42]. In this work, we also synthesized the uniform spherical TiO_2 particles with particle diameter of ca. 290 nm by adapting the previous synthetic method [42]. The SEM image (Figure 1b) clearly shows that uniform spherical particles with a white color powder were well synthesized. After the calcination step, amorphous TiO_2 could be crystallized to the crystalline counterparts, which consist of either anatase or rutile phases. The spherical morphology was well maintained, even after high temperature calcination, and the calcined sample showed white color, which is an intrinsic property of crystalline TiO_2 (Figure 1c). Chemical reduction using Li/EDA solution could selectively reduce the rutile phase of crystalline TiO_2. As previously reported by Park et al., Li-EDA as a strong reducing agent in a superbase can selectively reduce rutile TiO_2 to disorder the surface, resulting in black rutile TiO_2 having a small band-gap [39]. Since the calcined TiO_2 spherical particles had different crystalline ratios of anatase and rutile depending on the calcination temperature, the degree of surface disordering was highly influenced, resulting in the reduced spherical TiO_2 particle with different colors. In practical terms, the reduced TiO_2 sample showed a blue color with a uniform particle dimension (Figure 1d). After photo-deposition of Pt nanoparticle on the surface of the reduced TiO_2 sample, the sample could be used as a photocatalyst for photochemical hydrogen production under solar-light irradiation (Figure 1a).

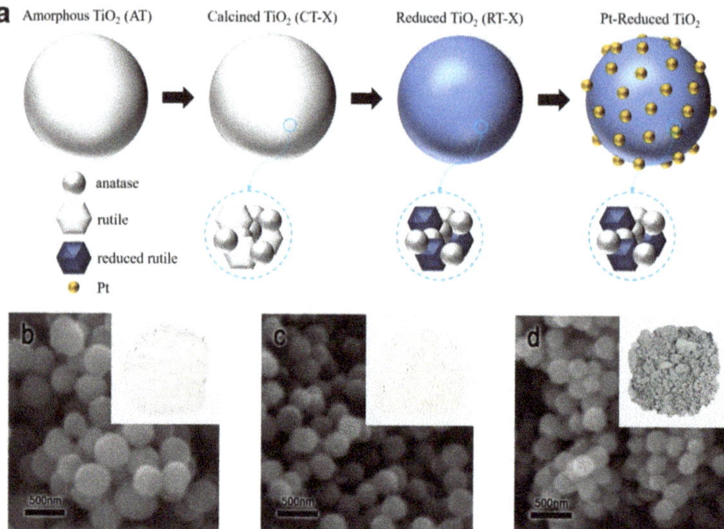

Figure 1. (**a**) Schematic illustration for synthesis of reduced TiO$_2$ samples (RT-x) and Pt-reduced TiO$_2$ catalyst. Corresponding SEM and digital images of (**b**) amorphous TiO$_2$ (AT), (**c**) calcined TiO$_2$ (CT-x), and (**d**) reduced TiO$_2$ (RT-x), respectively.

The morphologies of the calcined TiO$_2$ and the reduced TiO$_2$ samples are investigated by SEM. Figure 2a shows the CT-600 (Calcined TiO$_2$ at 600 °C) sample calcined at 600 °C, which reveals uniform spherical morphology with diameter of ca. (281 ± 31) nm. As the calcination temperature increases to 800 °C, the spherical morphology is well maintained, indicating the high thermal stability of the synthesized TiO$_2$ particles (Figure 2b). The average diameter of CT-800 is ca. (280 ± 37) nm, which is almost similar to that of CT-600. Although chemical reduction is carried out to produce the colored TiO$_2$ particles, the overall morphology with diameter is unchanged. All RT-x (Reduced TiO$_2$) samples show the monodispersed, spherical morphology with similar diameter compared to the mother CT sample, indicating high chemical stability (Figure 2c,d).

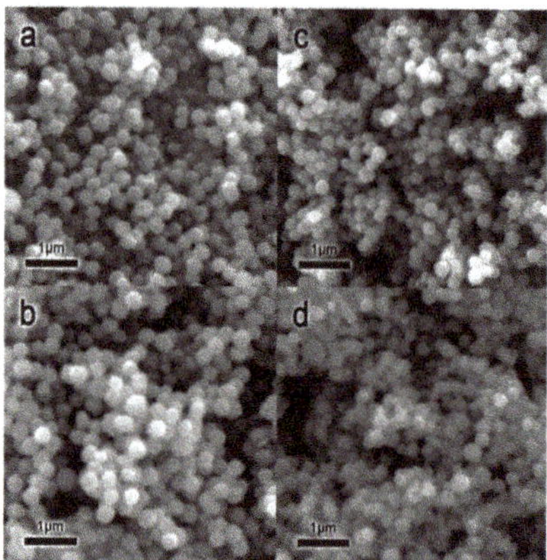

Figure 2. SEM images of (**a**) CT-600, (**b**) CT-800, (**c**) RT-600, and (**d**) RT-800. The letters CT and RT indicate calcined and reduced TiO$_2$, respectively. The number after CT and RT indicates its calcination temperature.

The crystalline characteristics of the calcined TiO_2 and the reduced TiO_2 samples are investigated by X-ray diffraction (XRD). Figure 3a shows the CT-500 sample revealed representative diffraction peaks of TiO_2 anatase phase at 2θ = (25.4°, 37.84°, 48.12°, 54.02°, 55.08°, and 62.68°), which are attributed to the (101), (004), (200), (105), (211), and (204) planes, respectively. As the calcination temperature increases, other peaks related to the TiO_2 rutile phase dramatically appear. CT-600 exhibits not only the sharp anatase peak, but also new rutile peaks at 2θ = (27.44°, 36.12°, 41.2°, 44°, 54.32°, and 56.6°), corresponding to the (110), (101), (111), (210), (211), and (220) planes, respectively. When the sample is calcined at ever higher temperatures of 700 °C (CT-700), the dominant rutile peaks become even sharper. As the calcination temperature increases further to 800 °C, major rutile peaks with a small trace of the anatase peak are observed. Based on the above results, it should be noted that the metastable anatase phase was continuously converted to the rutile phase by the thermal transformation of the TiO_2 crystalline phase.

The average composition of anatase to rutile phase was calculated from the relative peak area of anatase (101) and rutile (110) peaks, using the following equation [43]:

$$[A]/\% = 100 \times I_A/(I_A + 1.265 \times I_R) \tag{1}$$

where I_A and I_R correspond to the relative areas of the anatase (101) and rutile (110) peaks, respectively. Hence, the rutile content is [R] = 100 − [A]. The rutile contents of the CT-X samples were estimated to be approximately (0%, 67%, 88%, and 100%) for the CT-500, CT-600, CT-700, and CT-800, respectively.

Figure 3b also shows the XRD patterns of the reduced TiO_2 sample. RT-500 sample shows the anatase diffraction peaks that are identical XRD patterns to the CT-500 sample. This indicates that the crystalline properties of the RT-500 sample are well maintained, even after the chemical reduction process using Li/EDA solution. RT-600 sample shows similar mixed crystalline patterns of both the anatase phase and rutile phase, but it shows a slightly higher relative peak intensity of anatase, compared to that of CT-600. As the calcination temperature increases, the anatase peak intensity of RT-x samples is interestingly enhanced, compared to that of CT-x. RT-700 and RT-800 exhibit the more obvious anatase (101) peaks, compared to CT-700 and CT-800 samples. In practical terms, the rutile contents were calculated to be approximately (0%, 66%, 84%, and 93%) for the RT-500, RT-600, RT-700, and RT-800 samples, respectively. It should be noted that after Li/EDA reduction, the anatase ratio is obviously increased, while that of the rutile is slightly decreased. It was recently reported that the rutile phase could be selectively disordered by Li in superbase EDA solution [39]. Interestingly, Li/EDA solution as the reducing agent could reduce the ordered white rutile to the disordered black one, resulting in the colored TiO_2 with diminution of the rutile phase. In our study, we also observed similar phenomena by using our calcined TiO_2 samples, which can have either anatase or rutile phase. Since Li/EDA reducing solution can selectively disorder the surface of rutile phase in the calcined TiO_2 spherical particles, significant chemical reduction can induce a decrease of the rutile surface orderliness, resulting in both a decrease of the rutile peak intensity and an increase of the anatase peak, respectively. Thus, the relative content of the anatase phase of RT-x sample increases, compared to that of the CT-x one. However, because the Li/EDA solution cannot completely reduce all the rutile content, the majority of rutile phases on RT-600, RT-700, and RT-800 samples still remain, with obvious appearance of the anatase peaks.

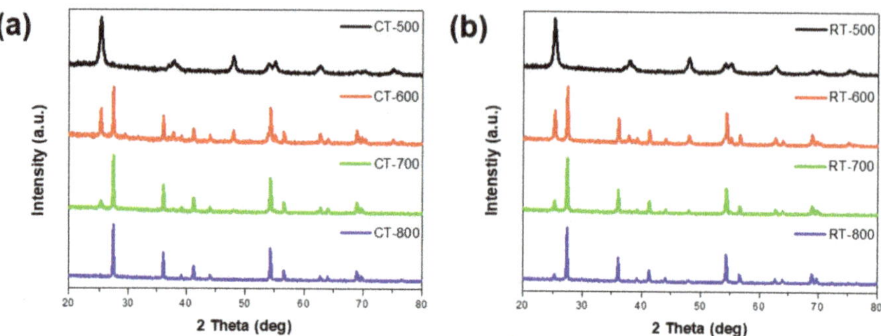

Figure 3. X-ray diffraction (XRD) patterns of (**a**) calcined TiO_2 sample (CT-x) and (**b**) reduced TiO_2 sample (RT-x).

The Raman spectra are also obtained to confirm the surface-structural changes of TiO_2 samples (Figure 4). As expected, the as-synthesized amorphous TiO_2 particle (AT) does not display any obvious peaks related to the crystalline structure. It showed small intensity changes at the Raman shift of ca. 1000 cm^{-1}, indicating the existence of a disordered surface [44]. CT-600 showed the obvious Raman peaks at Raman shift of ca. 400, 550, and 650 cm^{-1}, respectively, indicating representative ordered anatase characteristics. After the disordering process using the Li/EDA solution, RT-600 not only showed the similar Raman peaks at ca. 400, 550, and 650 cm^{-1}, but also obvious signal changes at ca. 1000 cm^{-1}. It should be noted that the disordered surface was formed by the Li/EDA reduction process. The RT-800 sample, which consists of mainly the rutile phase, displays a significant peak related to the disordered surface of rutile. The above features are also consistent with the trend of the XRD results. Based on the above XRD and Raman spectra, we conclude that the surface disordering of rutile can be easily achieved by a simple Li/EDA reduction process. Based on our observation, it can be considered that the original Ti^{4+} state in the boundary of the rutile crystalline grain is preferentially reduced to Ti^{3+} in superbase conditions in the presence of Ethylenediamine, which is a similar phenomenon to that previously reported [39].

We also confirmed if the disordered rutile surface could be recovered to the ordered surface by recalcining the RT-800 sample at 800 °C. As shown in Figure S1, CT-800 showed the obvious peaks at Raman shift of ca. 451 and 615 cm^{-1}, respectively, indicating the ordered rutile surfaces. After Li/EDA treatment, Raman peaks related with the ordered rutile surface of RT-800 were completely disappeared indicating the selective disordering of the rutile surface. When the sample is recalcined again at 800 °C, RT-800 recalcination exhibited the peaks of the ordered rutile surface again, indicating recovery of the disordered surface to the ordered counterpart. It should be noted that the ordered rutile surface can first be disordered through the Li/EDA reduction process, then the disordered surface can be recovered to the original ordered rutile surface by heat treatment. Based on our observation, it can be concluded that the original Ti^{4+} state in the boundary of the rutile crystalline grain is reduced to Ti^{3+}, then re-oxidized to Ti^{4+} states by sequential Li/EDA reduction and recalcination, respectively. It is consistent with the results of a previous study reported by Park et al. [39]. Since the disordered surface indicates different optical property, such as narrowed band-gap and enhanced absorption in visible light, it could be believed that our reduced TiO_2 sample showed different band-gap properties and enhanced photochemical performance under visible light conditions.

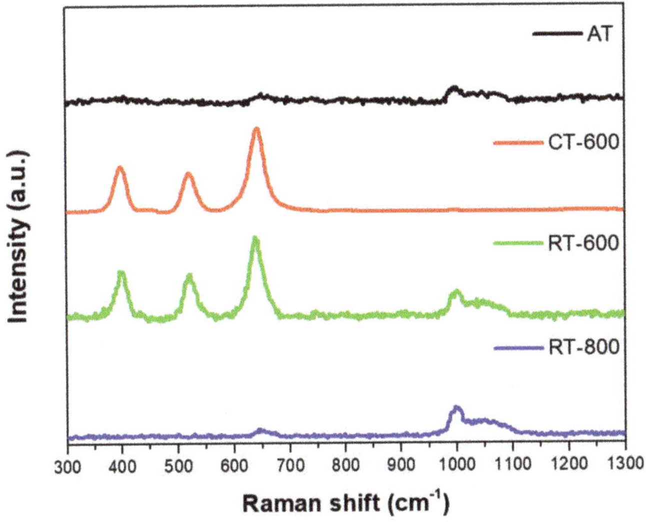

Figure 4. Raman spectra of amorphous TiO_2 (AT), CT-600, RT-600, and RT-800.

To investigate the light absorption ability and optical property of TiO_2 samples, we carried out UV-Vis diffuse reflectance spectroscopy (UV-Vis DRS). We obtained the absorption spectra of the TiO_2 samples employed in this work by UV-Vis DRS techniques (Figure 5). The as-synthesized amorphous TiO_2 sample (AT) can only absorb the UV region, indicating poor light absorbance towards solar light. While the calcined TiO_2 sample at 600 °C (CT-600) displays its main absorption of UV light until ca. 400 nm, the reduced TiO_2 sample (RT-600) shows the absorption edge red-shifted and significant increase of light absorption in the range ca. 400–650 nm, indicating large adsorption in the range of visible light. The RT-800 sample has even broader absorption in the range of visible light. We also estimated the band-gap energy of the above samples by using Tauc plot [45]. The band gap energy values of CT, CT-600, RT-600, and RT-800 are calculated as ca. 3.18, 2.93, 2.44, and 0.82 eV, respectively.

It is well known that pure TiO_2 shows a band-gap of ca. 3.0–3.2 eV [18]. In this work, the CT-600 sample, which has the mixed phase of anatase and rutile, displays a quite wide band-gap value of ca. 2.93 eV, even though there is small difference compared to previous results [10,11,18]. After chemical reduction using the Li/EDA solution, the RT-600 sample shows a narrowed band-gap (ca. 2.44 eV) with large absorption toward visible light. As previously reported, the Li/EDA solution can selectively make the disordered rutile surface of TiO_2 with a mixed phase of anatase and rutile [39]. Although undergoing the same Li/EDA reduction process, the anatase surface of the RT-500 sample can be well maintained, but some rutile surface of RT-600 can be selectively disordered. It is well known that the disordered surface can increase visible light absorption and utilize low-energy light on photocatalysis, which originates from either the regulated conduction band (CB) or valence band (VB) position, and indirect electron recombination [46]. In this study, we also observed that the reduced TiO_2 samples (RT-600 and RT-800) had enhanced light absorption ability and optical property, which should originate from the disordered surface of the rutile phase. Thus, during photocatalysis, our reduced TiO_2 samples should show enhanced performance.

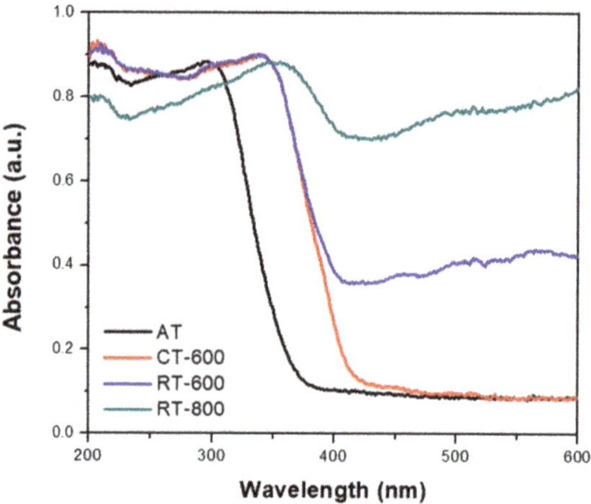

Figure 5. UV-Vis diffuse reflectance spectroscopy (UV-Vis DRS) spectra of amorphous TiO$_2$ (AT), CT-600, RT-600, and RT-800.

The photocatalytic hydrogen production activity of the TiO$_2$ sample was investigated under solar light (1.5 air mass, 1.5 AM) irradiation using methanol/H$_2$O solution (Figure 6a,b). Before employing the synthesized TiO$_2$ samples as photocatalysts, 1 wt.% of Pt was deposited on the surface of each TiO$_2$ sample. Figure 6a shows that when the Pt/CT-600 sample was used as the photocatalyst, only a negligible amount of hydrogen was produced. This indicates the low photocatalytic activity of Pt/CT-600, which is attributed to small photon absorption of the CT-600 sample toward solar light, which consists of mainly visible and infrared light. However, Pt/RT-600 shows remarkable improvement in photocatalytic hydrogen production (Figure 6a). It should be noted that both the improved visible light absorption and the narrowed band-gap of RT-600 enhance the light absorption, resulting in the dramatically improved photocatalysis activity of Pt/RT-600. Figure 6b shows the effect of calcination temperature of RT-x samples on photocatalytic hydrogen production with light irradiation time. Among the catalysts tested, the Pt/RT-600 catalyst shows the highest catalytic activity. The relative photocatalytic activity of the catalysts toward hydrogen production follows the order: Pt/RT-600 > Pt/RT-500 > Pt/RT-700 ≈ PT/RT-800.

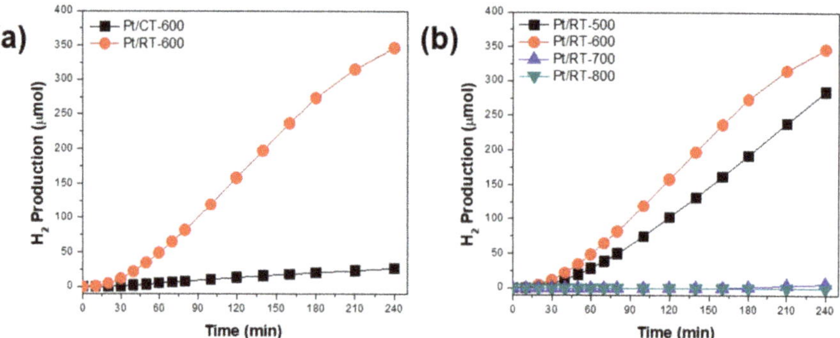

Figure 6. Hydrogen production amount of 1 wt % Pt-deposited TiO$_2$ photocatalyst: (**a**) CT-600 and RT-600; and (**b**) RT-500, RT-600, RT-700, and RT-800.

The photocatalytic activity of RT-x samples can be explained from the above characterization results. Since all RT-x samples was prepared from the same colloidal amorphous TiO_2 by calcination at different temperatures, followed by the same Li/EDA reduction process, it should be noted that the performance differences originate from the different degrees of rutile surface disordering, which is ascribed to different crystalline properties. The RT-500 sample, which is mainly anatase phase, has a small amount of disordered surface. In addition, since there is a small portion of UV-light in solar light, Pt/RT-500 can generate a considerable amount of hydrogen. The RT-600 sample, which consists of a mixed crystalline phase of anatase and rutile, can not only favorably absorb visible light over the reduced rutile, but it can also separate charge carriers to anatase, resulting in a lot of elongated charge carriers to accelerate photocatalysis. It is well known that the outstanding activity of P25-TiO_2 is mainly contributed by the mixed phase composition of anatase and rutile, which has beneficial effects on the absorption and separation of excited charges. Even though there is still controversy over the exact functions of each crystalline phase in the mixed phase, it is certain that the existence of the anatase–rutile mixed phase can have an unexpected and beneficial performance in photocatalysis [47,48]. Our RT-600 sample also had similar beneficial effects on light absorption and charge separation, due to the existence of the mixed phase. In practical terms, Pt/RT-600 shows the best hydrogen production performance. However, as the calcination temperature increases, the crystalline phase becomes mainly a rutile phase. RT-700 and RT-800 have a major rutile crystalline phase and a large portion of the disordered surface, resulting in the large absorption of visible light in the UV-DRS data. Although they can absorb visible light, the recombination of excited electron-hole pairs is severely constrained, due to the highly disordered surface and negligible portion of anatase. Although RT-800 sample can absorb the most visible light and the color of the catalyst is dark blue, Pt/RT-800 shows negligible hydrogen production.

To further confirm the photochemical properties of the TiO_2 samples, we measured the photocurrent by conducting chronoamperometry (CA) under an inducing potential of 0.6 V (vs. Ag/AgCl) and a periodic irradiation of solar light with a 400 nm cut-off filter. Figure 7 shows that the photocurrent is closely related to the degree of charge separation under light irradiation. Without light exposure, samples display electrochemical currents. When the catalysts are irradiated by light conditions, the current density can be increased due to the contribution from photo-generated electrons. The TiO_2 sample (CT-600) calcined at 600 °C exhibits a considerable photocurrent, indicating the existence of charge separation by light irradiation. The RT-600 sample shows the largest photocurrent among the TiO_2 samples tested. However, the RT-800 sample displays smaller photo-generated current than RT-600, indicating the restricted recombination of electron-holes, even though it can absorb a large portion of solar light. Based on both the characterization results and photoelectrochemical data, it can be concluded that the RT-600 photocatalyst, which consists of mixed phase with a disordered rutile surface, can have advantageous effects, such as visible light absorption, and favorable charge separation, resulting in improved photocatalysis activity on photocatalytic hydrogen production.

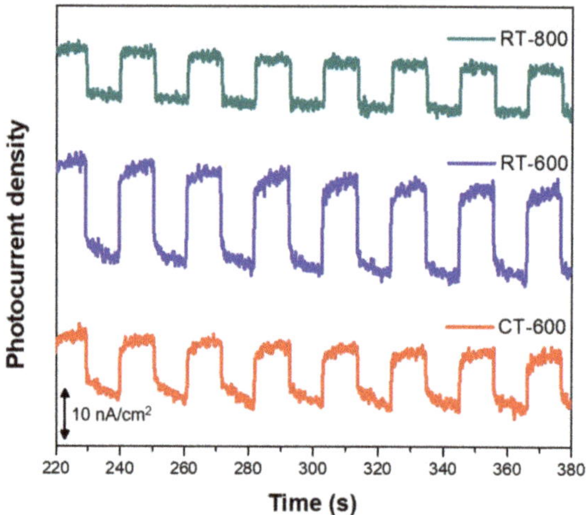

Figure 7. Photoelectrochemical chronoamperometry curves of various TiO_2 samples obtained at 0.6 V vs. Ag/AgCl under solar light illumination with 400 nm cut-off filter.

3. Materials and Methods

3.1. Materials

Ethyl alcohol (C_2H_5OH, 99.9%, anhydrous), Methyl alcohol (CH_3OH, 99.9%, anhydrous), acetonitrile (ACN, CH_3CN 99.9%, special guaranteed grade), and ammonium hydroxide (NH_4OH, 28%) were obtained from Daejung Chemical Company. Titanium (IV) n-butoxide (TBOT, 97%, reagent grade) and Hydroxypropyl cellulose (HPC, MW ≈ 80,000) were obtained from Aldrich. Metallic Li foil was purchased from Alfa Aesar chemical company. Ethylenediamine (EDA, 99%, guaranteed grade) was obtained from Sigma-Aldrich. All chemicals were used as received.

3.2. Synthesis

Spherical TiO_2 particles with average diameter of 290 nm were prepared by a sol-gel synthesis in mixed solvent, followed by calcination, as recently reported [42]. HPC (50 mg), as a surfactant, was completely dissolved in the mixed solvent solution (100 mL) of ethanol and acetonitrile with a volume ratio of 3:1. After completely dissolving the HPC, ammonia solution (0.8 mL) was added to the above solution. After stirring for 20 min, a mixture of TBOT (4 mL) in the mixed solvent of ethanol (12 mL) and acetonitrile (4 mL) was quickly injected into the above solution. The mixture was vigorously stirred for 2 h at room temperature. The white precipitate was isolated by centrifugation, and washed with ethanol and with de-ionized (D.I.) water several times. Then, the amorphous AT sample was obtained by drying under vacuum.

The dried AT sample was charged in an alumina boat in a furnace, and calcined at the desired temperature for 3 h under air conditions. The amorphous AT samples were crystallized to either anatase or rutile phase, which is highly dependent on the calcination temperature. Calcined CT samples are denoted as CT-X (where X is the calcination temperature). To synthesize colored TiO_2 spherical particle, we used the Li-EDA reduction method, which was recently developed by Park et al. [39]. A piece of metallic Li foil (45 mg) was dissolved in ethylenediamine (40 mL) to form a solvated electron solution. The calcined CT-X (400 mg) samples were added into the above solution, and vigorously stirred for 6 days under inert (N_2) conditions. After 6 days, a diluted HCl solution (1 M, 6.5 mL) was slowly added dropwise into the mixture, in order to quench the excess electron.

The reduced TiO_2 particles were isolated by centrifugation, washed with D.I. water 3 times, and dried in vacuum chamber at room temperature to give RT-X (where X is the calcination temperature).

3.3. Characterization

Digital photo images of TiO_2 samples were obtained by the digital camera function of iPhone (Apple Inc.). The particle morphology and uniformity were investigated by scanning electron microcopy (SEM, JSM-6060, JEOL). The crystalline properties of the TiO_2 samples were investigated by X-ray diffraction (XRD, D/MAX 2200, Rigaku). Optical absorbance spectra were studied by UV-vis spectroscopy using a UV-vis spectrophotometer with diffuse reflectance accessary (UV-DRS, V670, Jasco). Photoelectrochemical analyses were carried out using a conventional three-electrode system, with Ag/AgCl as a reference electrode, and Pt gauze as a counter electrode. The working electrode was prepared by deposition of a sample slurry on indium-tin oxide (ITO) glass (1 × 1 cm) [12]. An aqueous Na_2SO_4 (0.1 mol/L) solution containing methanol (10 vol.%) was used as the electrolyte. Chronoamperometry tests were conducted using a potentiostat (SP-150, BioLogic).

3.4. Photocatalytic Hydrogen Production

Photocatalytic hydrogen production was conducted in a Pyrex glass reactor. TiO_2 photocatalyst samples (20 mg) were well dispersed in an aqueous methanol solution (50%, 50 mL). An ABET 150 W Xe lamp (ABET technologies inc. USA) with an AM 1.5 G air mass filter was used as a light source for solar light irradiation. The amount of hydrogen produced was determined by conventional gas chromatography with a thermal conductivity detector (GC-TCD, HP-5890 equipped with a Molecular Sieve-5A packed column).

4. Conclusions

We synthesized the uniform spherical TiO_2 particles with disordered rutile surface, characterized both the physicochemical and optical properties, and demonstrated photocatalytic performances on photochemical hydrogen production under solar light conditions. The synthesis involves several sequential processes: (i) Synthesis of uniform-sized amorphous TiO_2 particles by sol-gel reaction of the TiO_2 precursor, (ii) calcination of amorphous TiO_2 sample to convert to its crystalline counterpart, and (iii) chemical reduction of the calcined TiO_2 sample to make the disordered rutile surface. The as-synthesized amorphous TiO_2 sample showed uniform and monodispersed spherical morphologies. Calcination at varied temperatures induced different crystalline characteristics, such as different ratios of anatase and rutile phases, and chemical reduction using Li/EDA enables selective disordering of the rutile surface, resulting in different optical characteristics, such as the degree of visible light absorption ability and band-gap energy. The chemically-reduced TiO_2 sample (RT-600) prepared by calcination at 600 °C, followed by Li/EDA reduction, displays beneficial characteristics in terms of light absorption and charge separation, such as a disordered rutile surface, and a mixed crystalline phase of anatase and rutile. Among the TiO_2 samples employed in this work, the RT-600 sample showed significantly enhanced catalytic activity in photocatalytic hydrogen production. We believe that the proposed technique reported in this study can provide an effective method for developing visible light-responsive TiO_2-based photocatalysts.

Supplementary Materials: The following are available online at http://www.mdpi.com/2073-4344/9/6/491/s1, Figure S1: Raman spectra of CT-800, RT-800, and RT-800 recalcination.

Author Contributions: Investigation, writing—original draft, N.Y.K.; investigation, H.K.L.; resources, J.T.M.; conceptualization, writing—review and editing, J.B.J.

Funding: This research was funded by Konkuk University in 2016.

Conflicts of Interest: The authors declare no conflicts of interest.

References

1. Rosen, M.A.; Koohi-Fayegh, S. The prospects for hydrogen as an energy carrier: an overview of hydrogen energy and hydrogen energy systems. *Energy Ecol. Environ.* **2016**, *1*, 10–29. [CrossRef]
2. Steinberg, M.; Cheng, H.C. Modern and prospective technologies for hydrogen production from fossil fuels. *Int. J. Hydrogen Energy* **1989**, *14*, 797–820. [CrossRef]
3. Zhang, T.; Amiridis, M.D. Hydrogen production via the direct cracking of methane over silica-supported nickel catalysts. *Appl. Catal. A* **1998**, *167*, 161–172. [CrossRef]
4. Ni, M.; Leung, D.Y.C.; Leung, M.K.H.; Sumathy, K. An overview of hydrogen production from biomass. *Fuel Process. Technol.* **2006**, *87*, 461–472. [CrossRef]
5. Wang, M.; Wang, Z.; Gong, X.; Guo, Z. The intensification technologies to water electrolysis for hydrogen production – A review. *Renew. Sustain. Energy Rev.* **2014**, *29*, 573–588. [CrossRef]
6. Yin, S.F.; Xu, B.Q.; Zhou, X.P.; Au, C.T. A mini-review on ammonia decomposition catalysts for on-site generation of hydrogen for fuel cell applications. *Appl. Catal. A* **2004**, *277*, 1–9. [CrossRef]
7. Khaselev, O.; Turner, J.A. A Monolithic Photovoltaic-Photoelectrochemical Device for Hydrogen Production via Water Splitting. *Science* **1998**, *280*, 425–427. [CrossRef] [PubMed]
8. Ni, M.; Leung, M.K.H.; Leung, D.Y.C.; Sumathy, K. A review and recent developments in photocatalytic water-splitting using TiO$_2$ for hydrogen production. *Renew. Sustain. Energy Rev.* **2007**, *11*, 401–425. [CrossRef]
9. Maeda, K.; Domen, K. Photocatalytic Water Splitting: Recent Progress and Future Challenges. *J. Phys. Chem. Lett.* **2010**, *1*, 2655–2661. [CrossRef]
10. Joo, J.B.; Dahl, M.; Li, N.; Zaera, F.; Yin, Y. Tailored synthesis of mesoporous TiO$_2$ hollow nanostructures for catalytic applications. *Energy Environ. Sci.* **2013**, *6*, 2082–2092. [CrossRef]
11. Joo, J.B.; Zhang, Q.; Dahl, M.; Zaera, F.; Yin, Y. Synthesis, crystallinity control, and photocatalysis of nanostructured titanium dioxide shells. *Int. J. Mater. Res.* **2013**, *28*, 362–368. [CrossRef]
12. Joo, J.B.; Zhang, Q.; Lee, I.; Dahl, M.; Zaera, F.; Yin, Y. Mesoporous Anatase Titania Hollow Nanostructures though Silica-Protected Calcination. *Adv. Funct. Mater.* **2012**, *22*, 166–174. [CrossRef]
13. Fujishima, A.; Honda, K. Electrochemical Photolysis of Water at a Semiconductor Electrode. *Nature* **1972**, *238*, 37–38. [CrossRef]
14. Li, W.; Wu, Z.; Wang, J.; Elzatahry, A.A.; Zhao, D. A Perspective on Mesoporous TiO$_2$ Materials. *Chem. Mater.* **2014**, *26*, 287–298. [CrossRef]
15. Yun, H.J.; Lee, H.; Joo, J.B.; Kim, W.; Yi, J. Influence of Aspect Ratio of TiO$_2$ Nanorods on the Photocatalytic Decomposition of Formic Acid. *J. Phys. Chem. C* **2009**, *113*, 3050–3055. [CrossRef]
16. Joo, J.B.; Liu, H.; Lee, Y.J.; Dahl, M.; Yu, H.; Zaera, F.; Yin, Y. Tailored synthesis of C@TiO$_2$ yolk–shell nanostructures for highly efficient photocatalysis. *Catal. Today* **2016**, *264*, 261–269. [CrossRef]
17. Joo, J.B.; Lee, I.; Dahl, M.; Moon, G.D.; Zaera, F.; Yin, Y. Controllable Synthesis of Mesoporous TiO$_2$ Hollow Shells: Toward an Efficient Photocatalyst. *Adv. Funct. Mater.* **2013**, *23*, 4246–4254. [CrossRef]
18. Scanlon, D.O.; Dunnill, C.W.; Buckeridge, J.; Shevlin, S.A.; Logsdail, A.J.; Woodley, S.M.; Catlow, R.A.; Powell, M.J.; Palgrave, R.G.; Parkin, I.P.; et al. Band alignment of rutile and anatase TiO$_2$. *Nat. Mater.* **2013**, *12*, 798–801. [CrossRef]
19. Pan, X.; Yang, M.-Q.; Fu, X.; Zhang, N.; Xu, Y.-J. Defective TiO$_2$ with oxygen vacancies: Synthesis, properties and photocatalytic applications. *Nanoscale* **2013**, *5*, 3601–3614. [CrossRef]
20. Liu, X.; Zhu, G.; Wang, X.; Yuan, X.; Lin, T.; Huang, F. Progress in Black Titania: A New Material for Advanced Photocatalysis. *Adv. Energy Mater.* **2016**, *6*, 1600452. [CrossRef]
21. Choi, W.; Termin, A.; Hoffmann, M.R. The Role of Metal Ion Dopants in Quantum-Sized TiO$_2$: Correlation between Photoreactivity and Charge Carrier Recombination Dynamics. *J. Phys. Chem.* **1994**, *98*, 13669–13679. [CrossRef]
22. Chen, L.-C.; Chou, T.-C. Photodecolorization of Methyl Orange Using Silver Ion Modified TiO$_2$ as Photocatalyst. *Ind. Eng. Chem. Res.* **1994**, *33*, 1436–1443. [CrossRef]
23. Asahi, R.; Morikawa, T.; Ohwaki, T.; Aoki, K.; Taga, Y. Visible-Light Photocatalysis in Nitrogen-Doped Titanium Oxides. *Science* **2001**, *293*, 269–271. [CrossRef] [PubMed]
24. Ohno, T.; Akiyoshi, M.; Umebayashi, T.; Asai, K.; Mitsui, T.; Matsumura, M. Preparation of S-doped TiO$_2$ photocatalysts and their photocatalytic activities under visible light. *Appl. Catal. A* **2004**, *265*, 115–121. [CrossRef]

25. Tang, Z.-R.; Zhang, Y.; Xu, Y.-J. Tuning the Optical Property and Photocatalytic Performance of Titanate Nanotube toward Selective Oxidation of Alcohols under Ambient Conditions. *ACS Appl. Mater. Interf.* **2012**, *4*, 1512–1520. [CrossRef] [PubMed]

26. Yun, H.J.; Lee, H.; Joo, J.B.; Kim, N.D.; Kang, M.Y.; Yi, J. Facile preparation of high performance visible light sensitive photo-catalysts. *Appl. Catal. B* **2010**, *94*, 241–247. [CrossRef]

27. Yun, H.J.; Lee, H.; Joo, J.B.; Kim, N.D.; Yi, J. Tuning the band-gap energy of $TiO_{2-x}C_x$ nanoparticle for high performance photo-catalyst. *Electrochem. Commun.* **2010**, *12*, 769–772. [CrossRef]

28. O'Regan, B.; Grätzel, M. A low-cost, high-efficiency solar cell based on dye-sensitized colloidal TiO_2 films. *Nature* **1991**, *353*, 737–740. [CrossRef]

29. Grätzel, M. Dye-sensitized solar cells. *J. Photochem. Photobiol. C* **2003**, *4*, 145–153. [CrossRef]

30. Tada, H.; Kiyonaga, T.; Naya, S.-i. Rational design and applications of highly efficient reaction systems photocatalyzed by noble metal nanoparticle-loaded titanium(iv) dioxide. *Chem. Soc. Rev.* **2009**, *38*, 1849–1858. [CrossRef]

31. Zhang, Q.; Lima, D.Q.; Lee, I.; Zaera, F.; Chi, M.; Yin, Y. A Highly Active Titanium Dioxide Based Visible-Light Photocatalyst with Nonmetal Doping and Plasmonic Metal Decoration. *Angew. Chem. Int. Ed.* **2011**, *50*, 7088–7092. [CrossRef] [PubMed]

32. Chen, X.; Liu, L.; Yu, P.Y.; Mao, S.S. Increasing Solar Absorption for Photocatalysis with Black Hydrogenated Titanium Dioxide Nanocrystals. *Science* **2011**, *331*, 746–750. [CrossRef] [PubMed]

33. Zhou, W.; Li, W.; Wang, J.-Q.; Qu, Y.; Yang, Y.; Xie, Y.; Zhang, K.; Wang, L.; Fu, H.; Zhao, D. Ordered Mesoporous Black TiO_2 as Highly Efficient Hydrogen Evolution Photocatalyst. *J. Am. Chem. Soc.* **2014**, *136*, 9280–9283. [CrossRef] [PubMed]

34. Yu, X.; Kim, B.; Kim, Y.K. Highly Enhanced Photoactivity of Anatase TiO_2 Nanocrystals by Controlled Hydrogenation-Induced Surface Defects. *ACS Catal.* **2013**, *3*, 2479–2486. [CrossRef]

35. Liu, N.; Schneider, C.; Freitag, D.; Hartmann, M.; Venkatesan, U.; Müller, J.; Spiecker, E.; Schmuki, P. Black TiO_2 Nanotubes: Cocatalyst-Free Open-Circuit Hydrogen Generation. *Nano Lett.* **2014**, *14*, 3309–3313. [CrossRef] [PubMed]

36. Wang, Z.; Yang, C.; Lin, T.; Yin, H.; Chen, P.; Wan, D.; Xu, F.; Huang, F.; Lin, J.; Xie, X.; et al. Visible-light photocatalytic, solar thermal and photoelectrochemical properties of aluminium-reduced black titania. *Energy Environ. Sci.* **2013**, *6*, 3007–3014. [CrossRef]

37. Sinhamahapatra, A.; Jeon, J.-P.; Yu, J.-S. A new approach to prepare highly active and stable black titania for visible light-assisted hydrogen production. *Energy Environ. Sci.* **2015**, *8*, 3539–3544. [CrossRef]

38. Razzaq, A.; Sinhamahapatra, A.; Kang, T.-H.; Grimes, C.A.; Yu, J.-S.; In, S.-I. Efficient solar light photoreduction of CO_2 to hydrocarbon fuels via magnesiothermally reduced TiO_2 photocatalyst. *Appl. Catal. B* **2017**, *215*, 28–35. [CrossRef]

39. Zhang, K.; Wang, L.; Kim, J.K.; Ma, M.; Veerappan, G.; Lee, C.-L.; Kong, K.-j.; Lee, H.; Park, J.H. An order/disorder/water junction system for highly efficient co-catalyst-free photocatalytic hydrogen generation. *Energy Environ. Sci.* **2016**, *9*, 499–503. [CrossRef]

40. Lu, H.; Zhao, B.; Pan, R.; Yao, J.; Qiu, J.; Luo, L.; Liu, Y. Safe and facile hydrogenation of commercial Degussa P25 at room temperature with enhanced photocatalytic activity. *RSC Adv.* **2014**, *4*, 1128–1132. [CrossRef]

41. Xing, M.; Zhang, J.; Chen, F.; Tian, B. An economic method to prepare vacuum activated photocatalysts with high photo-activities and photosensitivities. *Chem. Commun.* **2011**, *47*, 4947–4949. [CrossRef] [PubMed]

42. Moon, J.T.; Lee, S.K.; Joo, J.B. Controllable one-pot synthesis of uniform colloidal TiO_2 particles in a mixed solvent solution for photocatalysis. *Beilstein J. Nanotechnol.* **2018**, *9*, 1715–1727. [CrossRef] [PubMed]

43. Su, R.; Bechstein, R.; Sø, L.; Vang, R.T.; Sillassen, M.; Esbjörnsson, B.; Palmqvist, A.; Besenbacher, F. How the Anatase-to-Rutile Ratio Influences the Photoreactivity of TiO_2. *J. Phys. Chem. C* **2011**, *115*, 24287–24292. [CrossRef]

44. Xue, X.; Ji, W.; Mao, Z.; Mao, H.; Wang, Y.; Wang, X.; Ruan, W.; Zhao, B.; Lombardi, J.R. Raman Investigation of Nanosized TiO_2: Effect of Crystallite Size and Quantum Confinement. *J. Phys. Chem. C* **2012**, *116*, 8792–8797. [CrossRef]

45. Murphy, A.B. Band-gap determination from diffuse reflectance measurements of semiconductor films, and application to photoelectrochemical water-splitting. *Sol. Energy Mater. Sol. Cells* **2007**, *91*, 1326–1337. [CrossRef]

46. Zhang, J.; Zhou, P.; Liu, J.; Yu, J. New understanding of the difference of photocatalytic activity among anatase, rutile and brookite TiO_2. *Phys. Chem. Chem. Phys.* **2014**, *16*, 20382–20386. [CrossRef]
47. Bickley, R.I.; Gonzalez-Carreno, T.; Lees, J.S.; Palmisano, L.; Tilley, R.J.D. A structural investigation of titanium dioxide photocatalysts. *J. Solid State Chem.* **1991**, *92*, 178–190. [CrossRef]
48. Hurum, D.C.; Agrios, A.G.; Gray, K.A.; Rajh, T.; Thurnauer, M.C. Explaining the Enhanced Photocatalytic Activity of Degussa P25 Mixed-Phase TiO_2 Using EPR. *J. Phys. Chem. B* **2003**, *107*, 4545–4549. [CrossRef]

Article

Photoelectrochemical Hydrogen Evolution and CO_2 Reduction over MoS_2/Si and $MoSe_2$/Si Nanostructures by Combined Photoelectrochemical Deposition and Rapid-Thermal Annealing Process

Sungmin Hong, Choong Kyun Rhee and Youngku Sohn *

Department of Chemistry, Chungnam National University, Daejeon 34134, Korea; qwqe212@naver.com (S.H.); ckrhee@cnu.ac.kr (C.K.R.)
* Correspondence: youngkusohn@cnu.ac.kr; Tel.: +82-42-821-6548

Received: 17 April 2019; Accepted: 25 May 2019; Published: 28 May 2019

Abstract: Diverse methods have been employed to synthesize MoS_2 and $MoSe_2$ catalyst systems. Herein, a combined photoelectrochemical (PEC) deposition and rapid-thermal annealing process has first been employed to fabricate MoS_2 and $MoSe_2$ thin films on Si substrates. The newly developed transition-metal dichalcogenides were characterized by scanning electron microscopy, Raman spectroscopy and X-ray photoelectron spectroscopy. PEC hydrogen evolution reaction (HER) was demonstrated in an acidic condition to show a PEC catalytic performance order of MoO_x/Si < MoS_2/Si << $MoSe_2$/Si under the visible light-on condition. The HER activity (4.5 mA/cm^2 at −1.0 V vs Ag/AgCl) of $MoSe_2$/Si was increased by 4.8× compared with that under the dark condition. For CO_2 reduction, the PEC activity was observed to be in the order of MoS_2/Si < MoO_x/Si << $MoSe_2$/Si under the visible light-on condition. The reduction activity (0.127 mA/cm^2) of $MoSe_2$/Si was increased by 9.3× compared with that under the dark condition. The combined electrochemical deposition and rapid-thermal annealing method could be a very useful method for fabricating a thin film state catalytic system perusing hydrogen production and CO_2 energy conversion.

Keywords: MoS_2; $MoSe_2$; photoelectrochemical deposition; rapid-thermal annealing; hydrogen evolution; CO_2 reduction

---

1. Introduction

Transition-metal dichalcogenides (TMDCs) have widely been studied for applications to energy and environment such as hydrogen evolution and CO_2 reduction [1–10]. Especially, molybdenum disulfide and diselenide (MoS_2 and $MoSe_2$) materials with two-dimensional character have been synthesized using diverse synthesis methods for their applications [11]. Ye et al. employed a chemical vapor deposition (CVD) method to synthesize monolayer MoS_2 followed by oxygen plasma treatment or hydrogen annealing. They showed that hydrogen evolution reaction (HER) activity (e.g., onset potential and current density) was increased substantially by engineering the defects [12]. To increase HER activity of MoS_2 or $MoSe_2$, various defect engineering techniques have been employed, which include laser irradiation [13], ion irradiation [14,15] and NaClO chemical etching [16]. Li et al. examined various defect sites of MoS_2 such as edge sites, S vacancies and grain boundaries and showed that edge sites and S vacancies (with optimal vacancy density of 7–10%) were main HER active sites [17]. Chang et al. employed lithium molten salts to synthesize 2H- and 1T-MoS_2 monolayers at calcination temperatures of 400~600 °C and above 1000 °C, respectively [18]. They observed that metallic 1T-MoS_2 showed a higher HER activity than that of semiconducting 2H-MoS_2. Two step hydrothermal method was employed to synthesize 1T@2H-$MoSe_2$ nanosheets, which showed a higher HER activity [19].

Wang et al. prepared MoS_2 nanosheets by mechanical exfoliation, transferred onto a SiO_2 surface and made a HER device [20]. Afterwards, they showed that HER activity was increased by applying an extra positive electric field. Guo et al. prepared oxygen-incorporated MoS_2 sheets on graphene by a hydrothermal method [21] and showed that the active edge sites and conductivity were increased by oxygen-incorporation and the electrical transfer was increased by hybridization. Consequently, the HER activity was found to be substantially increased. Zhu et al. fabricated h-MoO_3/$1T$-MoS_2 heterostructures and tested photoelectrocatalytic HER activity to show better activity compared with those of $1T@2H$-MoS_2 and α-MoO_3/MoS_2 [22]. A two-step ($MoO_3 + H_2 \rightarrow MoO_2$ and $MoO_2 + Se$ vapor $\rightarrow MoSe_2$) chemical vapor deposition (CVD) process was employed to fabricate vertically aligned core–shell MoO_2/$MoSe_2$ nanosheet arrays which showed better HER activity than those of MoO_2 and $MoSe_2$ [23]. Electrochemical CO_2 reduction is another potential application area for MoS_2 and $MoSe_2$ [4,24–29]. Francis et al. tested a single crystal MoS_2 electrode for CO_2 reduction and showed a Faradaic efficiency of ~3.5% at -0.59 V (vs Reversible Hydrogen Electrode) for a reduction product of 1-propanol [4]. Asadi et al. reported that electrochemical CO_2 reduction product for vertically aligned MoS_2 in an ionic liquid was found to be CO with a CO_2 reduction current density of 130 mA cm^{-2} at -0.764 V [28,29].

Electrodeposition has popularly been employed for the cheap fabrication of thin films on a substrate, where major factors determining the nature of thin film include electrolyte, pH, deposition time and an applied potential [30]. For electrodeposition (under the dark condition) of Mo oxides on a substrate, some studies have been reported [31–34]. However, no studies have been reported for photoelectrochemical deposition of Mo oxides. Petrova et al. used Al substrates for electrodeposition of Mo oxides in Mo ion electrolyte ($Mo(NH_4)_6Mo_7O_{24}\cdot4H_2O$, 20 g/L) at pH of 8–10 adjusted by a NH_3-CH_3COONH_4 buffer [31]. Pd–MoO_x catalyst on glassy carbon electrode was reported to be fabricated by electrodeposition at potential ranges between -0.73 and $+0.2$ V in a mixed solution of 2 mM $PdCl_2$, 15 mM Na_2MoO_4 and 0.2 M HCl [32]. The dominant oxidation state of Mo was found to be +6. Uniform Mo oxide (+6, +5 and +4 oxidation states) nanostructure arrays (nanotubes at pH = 2.7 and nanowires at pH = 5.5) were prepared by a template electrodeposition in Mo ion electrolyte ($(NH_4)_6Mo_7O_{24}\cdot4H_2O$, 50 g/L) [33]. Electroless-photochemical deposition (PCD) has also been demonstrated for the preparation of metal sulfide thin films. Soundeswaran et al. prepared CdS films on indium tin oxide (ITO) glass using 1–10 mM $Cd(CH_3COO)_2$ and 100 mM $Na_2S_2O_3$ solution at pH = 3.0–4.5 under irradiation of ultraviolet (UV) light (100 mW/cm^2) [35]. For this reaction, S and electrons were initially formed by UV-excitation of $S_2O_3^{2-}$ ions and reacted with Cd metal ions to form CdS. Podder et al. prepared Cu_xS thin films on ITO glass using a similar method [36].

Herein, a new methodology of combined photoelectrochemical deposition ($Mo^{6+} + 6OH^- \rightarrow MoO_3 + 3H_2O$ accelerated by an enhanced photocurrent) and rapid-thermal annealing (RTA) sulfurization (or selenization) process ($2MoO_3 + 4S$ or $Se \rightarrow 2MoS_2$ or $MoSe_2 + 3O_2$) was introduced to fabricate thin MoS_2 and $MoSe_2$ films on Si substrates. HER and CO_2 reduction tests were demonstrated to show a potential applicability to energy and environment. A major advantage of the combined method is time-saving and cost effective. Another advantage is morphology and thickness-controlled by tuning applied voltage, deposition time and the electrolyte condition and so forth. Overall, the present developed method could be further improved and widely used for developing better thin film systems for diverse application areas.

2. Results

Surface morphology, crystal phase formation and surface chemical states were examined using scanning electron microscopy (SEM), Raman and X-ray photoelectron spectroscopy (XPS), respectively. Hydrogen evolution reaction (HER) and CO_2 reduction were tested using the three electrode system. The experimental results are described below.

2.1. SEM Morphology

Figure 1 shows the SEM images of MoO$_x$/Si, MoS$_2$/Si and MoSe$_2$/Si samples. For the as-photoelectrodeposited MoO$_x$/Si sample, larger (200~400 nm diameter) and smaller (< 50 nm) nanoparticles were formed on the Si surface. Two different particle size distributions may be due to mixed Volmer-Weber island growth mode and Ostwald ripening process. The oxidation states were confirmed by XPS, discussed below. Under the dark condition, no electrodeposition was observed. For the SEM image of MoS$_2$ formed by RTA process of MoO$_x$/Si sample, two different size distributions were also observed for the sample as expected. For the SEM image of MoSe$_2$ formed by RTA process of MoO$_x$/Si sample, the surface morphology showed more uniform nanostructure.

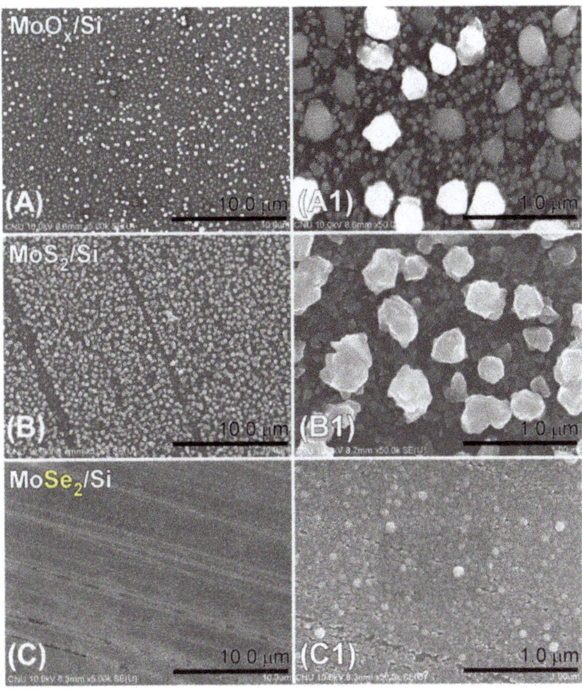

Figure 1. Scanning electron microscope (SEM) images of MoO$_x$/Si (**A** and **A1**), MoS$_2$/Si (**B** and **B1**) and MoSe$_2$/Si (**C** and **C1**) samples.

2.2. Raman Spectroscopy

Raman spectra (Figure 2) were obtained to examine the detailed crystal phase formation for the catalyst systems. For all the samples, the strongest peak was commonly observed at 524 cm^{-1} (not shown in the Figure), attributed to Si used as a support [37]. A weak peak at ~300 cm^{-1} for MoO$_x$/Si was due to the phonon mode of Si [37]. No Raman peaks of Mo oxides were observed, indicating that the as-photoelectrodeposited sample was ultrathin and/or amorphous. For MoS$_2$ sample (Figure 2B), two Raman peaks were observed at 385 and 411 cm^{-1}, attributed to in-plane Mo-S (E$_{2g}$) and out-of-plane Mo-S (A$_{1g}$) vibration modes of hexagonal MoS$_2$ [38,39]. For MoSe$_2$ sample (Figure 2C), a strong Raman peak was observed at 238 cm^{-1}, attributed to Mo-Se A$_{1g}$ mode. Other two peaks at 169 and 286 cm^{-1} were assigned to E$_{1g}$ and E$_{2g}$ modes, respectively [40,41]. The A$_{1g}$ mode was found to be stronger than the E$_{2g}$ mode. Overall, based on the Raman data, the RTA process was found to be efficient for the fabrication of MoS$_2$ and MoSe$_2$.

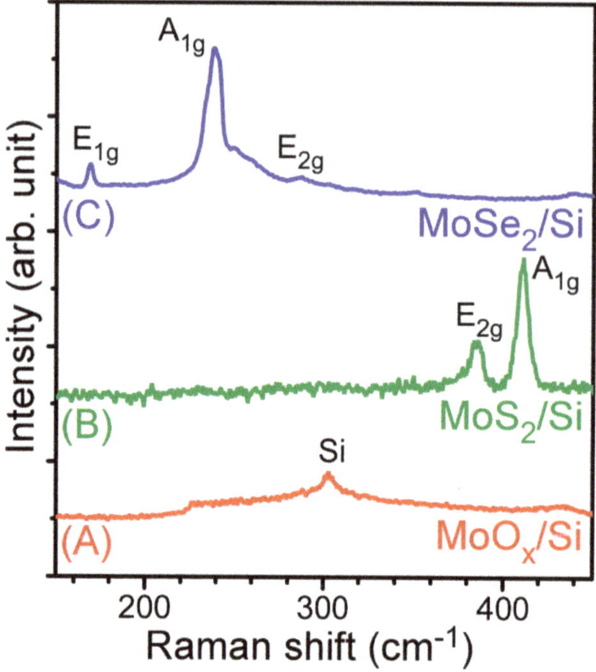

Figure 2. Raman spectra of (A) MoO_x/Si, (B) MoS_2/Si and (C) $MoSe_2/Si$ samples.

2.3. X-ray Photoelectron Spectroscopy

X-ray photoelectron spectra (XPS) of the three catalyst systems are displayed in Figure 3. For the as-photoelectrodeposited MoO_x/Si, Mo and O elements were dominantly observed. Mo $3d_{3/2}$ and $3d_{5/2}$ XPS peaks were observed at 235.5 and 232.5 eV, respectively with a spin-orbit splitting energy of 3.0 eV. This is attributed to Mo^{6+} oxidation state of MoO_3 [42]. Furthermore, the other Mo $3d_{3/2}$ and $3d_{5/2}$ XPS peaks were observed at 233.8 and 230.8 eV, respectively with a spin-orbit splitting energy of 3.0 eV. This could be due to Mo^{5+} oxidation state. Based on the Mo 3d XPS fitting, 42% and 52% of Mo were 6+ and 5+ oxidation states, respectively. The corresponding broad O 1s peak was observed at 530.6 eV, attributed to the lattice oxygen of Mo oxides (MoO_x). For the Mo 3d XPS of the MoS_2/Si sample, Mo $3d_{3/2}$ and $3d_{5/2}$ XPS peaks were observed at 232.2 and 229.0 eV, respectively with a spin-orbit splitting energy of 3.2 eV. This is attributed to Mo^{4+} oxidation state of MoS_2 [43,44]. A small peak at 226.3 eV was due to S 2s [45]. The corresponding S $2p_{1/2}$ and S $2p_{3/2}$ XPS peaks were observed at 163.1 eV and 161.9 eV, respectively with a spin-orbit splitting of 1.2 eV. This is in good agreement with the literature reported by Jian et al. for 1T phase MoS_2 [44]. For the Mo 3d XPS of the $MoSe_2/Si$ sample, Mo $3d_{3/2}$ and $3d_{5/2}$ peaks were observed at 231.7 and 228.6 eV, respectively with a spin-orbit splitting energy of 3.1 eV. This is in good agreement with +4 oxidation state of $MoSe_2$ [5]. The corresponding Se $3d_{5/2}$ and $3d_{3/2}$ peaks were found to be located 55.1 and 54.2 eV, respectively. The O 1s XPS peak was observed at 533.1 eV, attributed adsorbed surface oxygen [42]. For MoO_x/Si sample, the Mo:O ratio was estimated to be 1:4.1. The overestimated oxygen was plausibly due to surface oxygen such as H_2O and OH. Mo:O ratio was estimated to be 1: 2.3 by only considering the lattice O 1s signal. For MoS_2/Si sample, Mo:S ratio was calculated to be 1:1.95, very close to MoS_2. For $MoSe_2/Si$ sample, Mo:Se ratio was calculated to be 1:2.15, close to $MoSe_2$. For the valence band (VB) XPS spectra, the density of states (DOS) near the Fermi level was more discernible for MoS_2 and $MoSe_2$ reflecting metallic/semiconducting states.

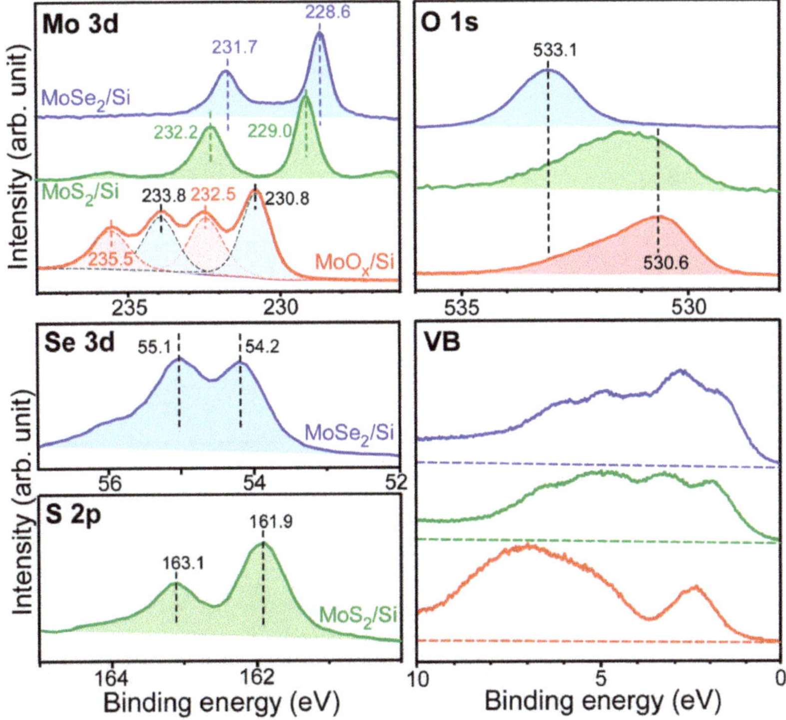

Figure 3. Mo 3d, O 1s, Se 3d, S 2p and valence band (VB) X-ray photoelectron spectra (XPS) for MoO_x/Si, MoS_2/Si and $MoSe_2$/Si samples.

2.4. Photoelectrochemcial Hydrogen Evolution

Three different catalyst systems of MoO_x/Si, MoS_2/Si and $MoSe_2$/Si were tested for hydrogen evolution reaction in 0.1 M H_2SO_4 electrolyte. Linear sweep voltammetry (LSV) curves (Figure 4) were obtained from +0.2 V to −1.0 V at a scan rate of 10 mV/sec after full nitrogen gas purging under the dark and visible light exposure conditions. Under the dark condition, the current density at -1.0 V (vs Ag/AgCl) showed the order of MoO_x/Si (0.08 mA/cm^2) < MoS_2/Si (0.11 mA/cm^2) << $MoSe_2$/Si (0.86 mA/cm^2) while the order changed to MoS_2/Si (0.51 mA/cm^2) < MoO_x/Si (0.64 mA/cm^2) << $MoSe_2$/Si (4.3 mA/cm^2) under the visible light exposure condition. They all commonly showed an increase in HER CD under visible light exposure. The inset in Figure 4 displays LSV curves taken under the light ON-and-OFF condition during the scan. It clearly showed that all three samples have photocatalytic activity.

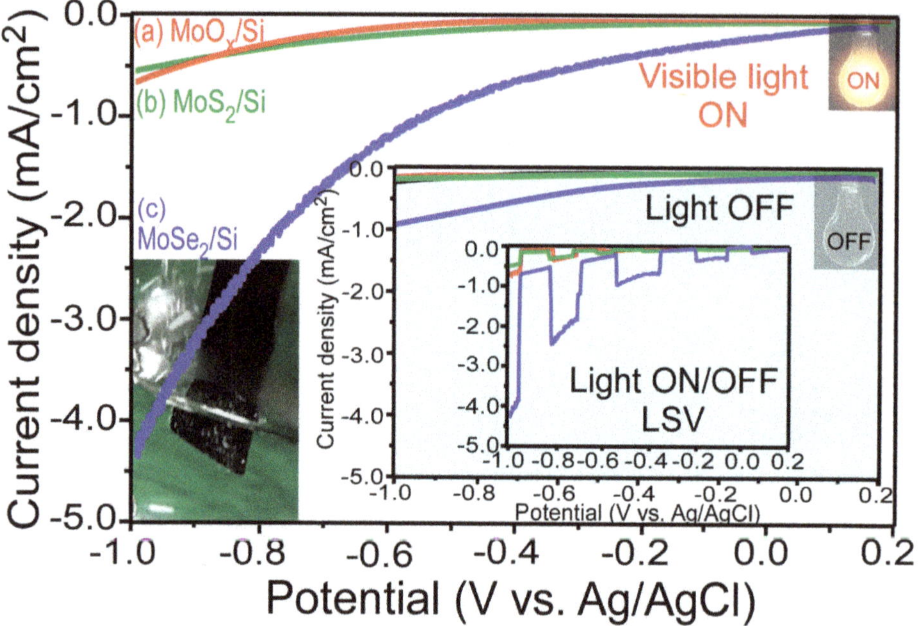

Figure 4. Linear sweep voltammetry curves (voltage range: +0.2~1.0 V) at a scan rate of 10 mV/sec) under visible light exposure for (a) MoO$_x$/Si, (b) MoS$_2$/Si and (c) MoSe$_2$/Si samples. Inset LSV curves were taken under the dark condition. The corresponding light ON-and-OFF LSV curves are displayed in the inset. Inset photo shows the bubble formed on the catalyst surface during LSV.

2.5. Photoelectrochemcial CO$_2$ Reduction

For CO$_2$ reduction in 0.1 M NaHCO$_3$ electrolyte, the LSV measurements (Figure 5) were obtained from +0.2 V to −1.0 V upon full N$_2$ and CO$_2$ gas purging at a scan rate of 10 mV/sec under the dark and visible light exposure conditions, respectively. Upon full N$_2$ gas purging in the electrolyte under dark condition (inset in Figure 5A), the current densities (CDs) at −1.0 V were observed to be 0.0043, 0.0041 and 0.012 mA/cm^2 for MoO$_x$/Si, MoS$_2$/Si and MoSe$_2$/Si, respectively. Under the visible light exposure condition (Figure 5A), the CDs were found to be drastically increased to 0.037, 0.011 and 0.018 mA/cm^2 for MoO$_x$/Si, MoS$_2$/Si and MoSe$_2$/Si, respectively. Upon full CO$_2$ gas purging in the electrolyte under the dark condition (inset in Figure 5B), the CDs at −1.0 V were found to be 0.007, 0.009 and 0.014 mA/cm^2 for MoO$_x$/Si, MoS$_2$/Si and MoSe$_2$/Si, respectively. Under the visible light exposure condition (Figure 5B), the CDs were found to be increased to 0.043, 0.023 and 0.13 mA/cm^2 for MoO$_x$/Si, MoS$_2$/Si and MoSe$_2$/Si, respectively. The CDs were increased by 6.0, 2.6 and 9.4×, respectively upon visible light exposure. The inset in Figure 5B displays LSV curves taken under the light ON-and-OFF condition during the scan. It clearly showed that all three samples also have photocatalytic activity.

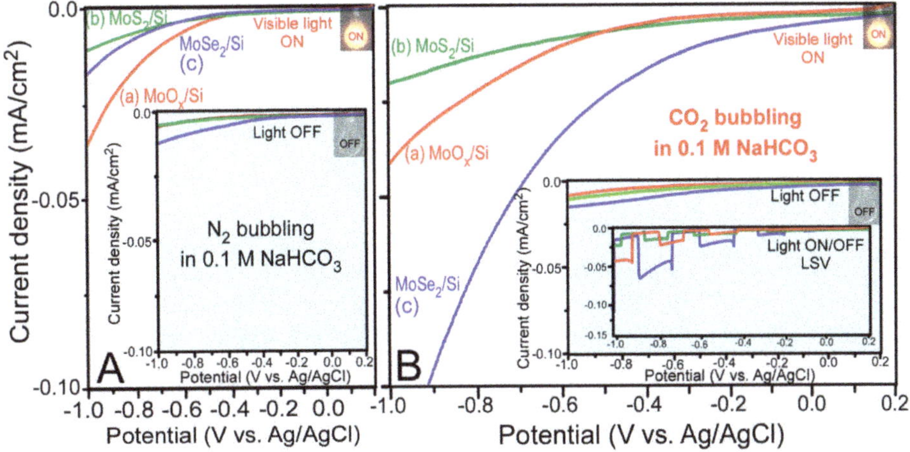

Figure 5. Linear sweep voltammetry curves (voltage range: +0.2~1.0 V) in N_2- (**A**) and CO_2-purged (**B**) 0.1 M NaHCO$_3$ electrolyte at a scan rate of 10 mV/sec under dark (in the corresponding inset Figure) and the visible light exposure condition for (a) MoO$_x$/Si, (b) MoS$_2$/Si and (c) MoSe$_2$/Si samples. The inset (**B**) shows the light ON-and-OFF LSV curves.

3. Discussion

The photo-electrochemical deposition method was first successfully employed to fabricate MoO$_x$ on a Si support. Upadhyay et al. reported MoO$_x$ nanoparticles on a stainless steel support prepared by electrodeposition at -1.0 V in a mixed solution of 0.1 M Na$_2$MoO$_4$ and 0.1 M NH$_4$NO$_3$ [46]. They reported oxidation states of Mo^{6+} and Mo^{5+} for MoO$_x$ and the corresponding O 1s peak at 530.6 eV. This is in good agreement with our present results. Liu et al. performed electrodeposition of MoO$_x$ films on a Ti (130 nm)/Si substrate in a mixed solution of 0.1 M Na$_2$MoO$_4$, 0.1 M Na$_2$EDTA and 0.1 M CH$_3$COONH$_4$ [47]. They concluded that the as-electrodeposited MoO$_x$ film (with oxidation states of Mo^{4+} and Mo^{5+}) was amorphous. This is also in good agreement with the present result, as discussed above (Raman spectra in Figure 2). Based on the SEM image, the photoelectrochemical deposition of MoO$_x$ on a Si support appeared to be occurred through Volmer-Weber island growth process [48]. Small islands were initially formed on the entire surface and then larger islands subsequently were grown. For the SEM images of MoS$_2$ and MoSe$_2$ by the RTA process, the morphology of MoS$_2$ was more similar to that of MoO$_x$, compared with that of MoSe$_2$. This indicates that less energy was required for the formation of MoS$_2$, compared with that for MoSe$_2$. Overall, the RTA process was found to be efficient for the formation of MoS$_2$ and MoSe$_2$ without much impacting the original morphology of electrodeposited MoO$_x$.

For HER in 0.1 M H$_2$SO$_4$, MoSe$_2$/Si showed a much higher electrochemical activity (or current density) than MoS$_2$/Si. For HER mechanism in the acidic condition, adsorbed hydrogen is known to be formed via $H_3O^+ + e^- \rightarrow H_{ad} + H_2O$. Then, hydrogen is generated via $H_{ad} + H_3O^+ + e^- \rightarrow H_2 + H_2O$ or $H_{ad} + H_{ad} \rightarrow H_2$ [23]. In the mechanism, hydrogen adsorption Gibbs free energy, ΔG_{HX} (X = S or Se) is known to play a major role in determining the activity [10]. The optimal condition is $\Delta G_{HX} = 0$ eV and ΔG_{HX} of MoSe$_2$ is closer to the optimal condition than that of MoS$_2$. The HER activity of MoSe$_2$ has commonly been reported to be higher than that of MoS$_2$ [49]. This is in good consistent with the present result. For HER of nanoflowers-like MoS$_2$ and MoSe$_2$ materials on GC electrodes in 0.5 M H$_2$SO$_4$, Ravikumar et al. reported that the activity (11 mA/cm^2 at 0.3 V vs RHE) of MoSe$_2$ showed a higher than that (7 mA/cm^2 at 0.3 V vs RHE) of MoS$_2$, attributed to higher electrical properties and defects [50], in good consistent with the present result. Evidently, based on the DOS near the Fermi level as discussed in Figure 3, the enhanced electronic conductivity could play an important role in HER and CO_2 reduction performances [22]. Because the Gibbs free energy is also known to be dependent on morphology (e.g., defects and edge sites) the catalyst fabrication methods is important for improving a catalyst activity. Upon visible light exposure on the catalyst surface during the LSV, an increased CD was commonly been observed. The enhancement factors (light ON/OFF CD ratio) for HER at −1.0 V upon visible light irradiation were observed to be 8.0, 4.7 and 5.0 for MoO_x/Si, MoS_2/Si and $MoSe_2$/Si, respectively (Figure 6). The photocatalytic HER activity under visible light was due to photo-generated electron by absorption of light in the visible region [22,42]. As mentioned above, in HER mechanism electron plays a crucial role in generation of hydrogen. Overall, the photocatalytic activity is all dependent on material nature (e.g., electrical conductivity), morphology, surface natures (e.g., defects) and light absorption efficiency. For the splitting of water, the molar stoichiometric ratio of H_2/O_2 is ideally 2.0 [51] assuming that no other side electrochemical reactions are involved for MoS_2 and $MoSe_2$ [1]. Before further discussion, it should be here mentioned that our conclusion was based only on the CD. The H_2/O_2 production ratios and CO_2 reduction products are needed to be further examined by gas chromatography [51].

For electrochemical CO_2 reduction, the current density was commonly been increased in CO_2-purged 0.1 M $NaHCO_3$ electrolyte, compared with that in N_2-purged 0.1 M $NaHCO_3$ electrolyte. This indicates that CO_2 reduction was occurred for all the samples. For CO_2 reduction (Figure 6), under the dark condition, the enhancement factors before (only N_2 bubbling) and after CO_2 bubbling were observed to be 1.6, 2.2, 1.2 for MoO_x/Si, MoS_2/Si and $MoSe_2$/Si, respectively. Under the visible light-on condition, the enhancement factors before (only N_2 bubbling) and after CO_2 bubbling were observed to be 1.1, 2.1 and 7.2 for MoO_x/Si, MoS_2/Si and $MoSe_2$/Si, respectively. The CDs upon light exposure were increased by 6.0, 2.6 and 9.4× for MoO_x/Si, MoS_2/Si and $MoSe_2$/Si, respectively. Overall, $MoSe_2$ showed the most dramatic photo-electrochemical CO_2 reduction efficiency. For $MoSe_2$, photogenerated electrons are created by light absorption in the visible region and electron-hole recombination is suppressed by good electron transport [1]. Consequently, the CD by photoelectrochemical CO_2 reduction is enhanced.

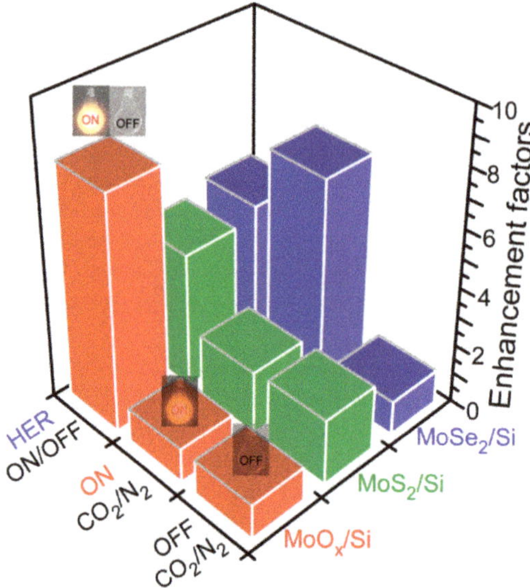

Figure 6. Enhancement factors for HER CD$_{light ON}$/ HER CD$_{light OFF}$, CD$_{CO2 bubbling}$/CD$_{N2 bubbling}$ under light ON condition and CD$_{CO2 bubbling}$/CD$_{N2 bubbling}$ under the light OFF condition for (**a**) MoO_x/Si, (**b**) MoS_2/Si and (**c**) $MoSe_2$/Si samples.

4. Materials and Methods

For photoelectrochemical Mo deposition (MoO_x/Si), a three-electrode (Ag/AgCl reference, Pt wire counter and Si working electrode) electrochemical cell was used using a VersaSTAT3 (Princeton Applied Research, Oak Ridge, TN, USA) potentiostat galvanostat. For the preparation of a Si working support electrode before electrodeposition, a single-side polished Si (100) wafer (B-doped p-type, thickness of 525 ± 20 μm, resistivity of 1–10 Ω·cm, 2 cm × 0.5 cm) was used as the support, cleaned in 2% HF solution to remove oxide layer and washed with deionized water by sonication. The electrolyte was a mix of 15 mM Na_2MoO_4 (99.0% extra pure, Samchun Chem. Co., Seoul, Republic of Korea), 1.0 M NaCl and 1.0 M NH_3Cl, where pH was adjusted to 9.2 using NH_3OH (28~30%, Samchun Chem. Co., Seoul, Republic of Korea). The photoelectrochemical deposition was performed at an applied potential of −1.5V (versus Ag/AgCl electrode) for 10 sec under visible light exposure onto the working electrode. In the present study, we only showed the samples prepared at fixed parameters among different applied potentials and deposition times. No efficient (or less uniform) Mo deposition occurred under the dark condition although the applied voltages and deposition times were varied. CDs of −0.10 and −0.92 mA/cm^2 were measured at −1.5 V under dark and visible light, respectively. The CD was enhanced by 9–10× in the potential ranges from −1.0 to −2.0 V. For the preparation of MoS_2 and $MoSe_2$ on Si support, a rapid-thermal annealing (RTA) method was employed using a LABSYS RTP-1200 (Nextron Co., Ltd., Busan, Republic of Korea). For this, a MoO_x/Si substrate was placed on a quartz plate (15 mm × 20 mm) in the RTA chamber. Sulfur (S.P.C. GR reagent, Shinyo Pure Chem. Co., Ltd., Hyogo, Japan) or Selenium (99.5+%, 100mesh, Sigma-Aldrich, St. Louis, MO, USA) powder (~0.02 g) was placed below the substrate. The chamber was maintained in 5% H_2 in He balance. The temperature heating rate was 12 °C/sec and the time was 6 sec at the maximum temperature of 700 °C. The surface morphology of the prepared samples was examined using a Hitachi S-4800 (Tokyo, Japan) scanning electron microscope at an electron acceleration voltage of 10.0 kV. Raman spectra with 514 nm laser line were obtained using a LabRAM HR-800 microRaman spectrometer (HORIBA Jobin Yvon, Kyoto, Japan). X-ray photoelectron spectra were taken using a Thermo Scientific K-Alpha$^+$ X-ray photoelectron spectrometer with micro-focused monochromatic Al Kα X-ray source and a hemispherical energy analyzer. XPS spectra curve fitting was performed using a XPSPEAK ver. 4.1 software. For XPS element quantification, XPS sensitivity factors of 2.75, 0.66, 0.54, and 0.67 were used for Mo 3d, O 1s, S 2p and Se 3d, respectively [52]. For photoelectrochemical HER and CO_2 reduction, a three-electrode electrochemical cell was also used using a VersaSTAT3 potentiostat/galvanostat. For HER, nitrogen gas was fully purged into the electrolyte (0.1 M H_2SO_4 solution) to minimize an effect of dissolved oxygen. Linear sweep voltammetry (LSV) was carried out at a scan rate of 10 mV s^{-1} from +0.2 V to −1.0 V under dark and visible light exposure conditions. A white LED USB Flashlight (A-10, Teckmedia) was used for visible light (400~700 nm) [53]. For CO_2 reduction experiment in 0.1 M $NaHCO_3$ solution, LSV was conducted after N_2 gas purging at a scan rate of 10 mV s^{-1} from +0.2 V to −1.0 V under dark and visible light irradiation conditions. The same LSV experiment was also conducted after CO_2 gas purging into the electrolyte to examine the CO_2 effect.

5. Conclusions

In this work, a combined photoelectrochemical deposition and rapid-thermal annealing method was first been employed to fabricate MoS_2 and $MoSe_2$ thin films on Si substrates. Photoelectrochemical HER and CO_2 reduction were demonstrated for the newly developed catalytic systems. The main results are as follows:

1. Mo oxides were successfully electrodeposited on a Si support under visible light exposure. Under the dark condition, the electrochemical deposition was less efficient.
2. A rapid-thermal annealing method was successfully introduced for the electrodeposited MoO_x/Si to fabricate MoS_2/Si and $MoSe_2/Si$ catalyst systems. Other impurity phases were not detected

mainly based on the Raman and XPS results. The maximum temperature was achieved by rapid heating to 700 °C of S or Se powers on the MoO_x/Si and maintained for 6 sec.

3. HER tests in 0.1 M H_2SO_4 electrolyte showed a catalytic activity order of MoO_x/Si < MoS_2/Si << $MoSe_2$/Si under dark and visible light-on conditions. The HER activity (4.5 mA/cm^2 at −1.0 V vs Ag/AgCl) of $MoSe_2$/Si was increased by 4.8× compared with that under the dark condition.

4. CO_2 reduction tests in 0.1 M $NaHCO_3$ electrolyte showed a catalytic activity order of MoO_x/Si, MoS_2/Si << $MoSe_2$/Si and MoS_2/Si < MoO_x/Si << $MoSe_2$/Si under dark and visible light-on conditions, respectively. The reduction activity (0.127 mA/cm^2) of $MoSe_2$/Si was increased by 9.3× compared with that under the dark condition.

The newly developed catalyst preparation method could be very useful for developing thin film catalyst systems for diverse application areas.

Author Contributions: Y.S. designed the experiments and wrote the paper; S.H. performed the experiments; C.K.R. provided valuable idea for obtaining the data.

Funding: This research was funded by the National Research Foundation of Korea (NRF) grant funded by the Korean government (MEST), grant number NRF-2016R1D1A3B04930123. The APC was funded by the National Research Foundation of Korea (NRF).

Acknowledgments: This work was financially supported by the National Research Foundation of Korea (NRF) grant funded by the Korean government (MEST) (NRF-2016R1D1A3B04930123)

Conflicts of Interest: The authors declare no conflict of interest.

References

1.   Yang, L.; Liu, P.; Li, J.; Xiang, B. Two-dimensional material molybdenum disulfides as electrocatalysts for hydrogen evolution. *Catalysts* **2017**, *7*, 285. [CrossRef]

2.   Theerthagiri, J.; Senthil, R.A.; Senthilkumar, B.; Polu, A.R.; Madhavan, J.; Ashokkumar, M. Recent advances in MoS_2 nanostructured materials for energy and environmental applications—A review. *J. Sol. Stat. Chem.* **2017**, *252*, 43–71. [CrossRef]

3.   Eftekhari, A. Molybdenum diselenide ($MoSe_2$) for energy storage, catalysis and optoelectronics. *Appl. Mater. Today* **2017**, *8*, 1–17. [CrossRef]

4.   Francis, S.A.; Velazquez, J.M.; Ferrer, I.M.; Torelli, D.A.; Guevarra, D.; McDowell, M.T.; Sun, K.; Zhou, X.; Saadi, F.H.; John, J.; et al. Reduction of aqueous CO_2 to 1-propanol at MoS_2 Electrodes. *Chem. Mater.* **2018**, *30*, 4902–4908. [CrossRef]

5.   Guo, W.; Chen, Y.; Wang, L.; Xu, J.; Zeng, D.; Peng, D.L. Colloidal synthesis of $MoSe_2$ nanonetworks and nanoflowers with efficient electrocatalytic hydrogen-evolution activity. *Electrochim. Acta* **2017**, *231*, 69–76.

6.   Guo, H.; Jin, C.; Liu, X.; Liu, J.; Vasileff, A.; Jiao, Y.; Zheng, Y.; Qiao, S. Emerging Two-dimensional nanomaterials for electrocatalysis. *Chem. Rev.* **2018**, *118*, 6337–6408.

7.   Mao, J.; Wang, Y.; Zheng, Z.; Deng, D. The rise of two-dimensional MoS_2 for catalysis. *Front. Phys.* **2018**, *13*, 138118. [CrossRef]

8.   Sun, Z.; Ma, T.; Tao, H.; Fan, Q.; Han, B. Fundamentals and challenges of electrochemical CO_2 reduction using two-dimensional materials. *Chem* **2017**, *3*, 560–587. [CrossRef]

9.   Gupta, U.; Rao, C.N.R. Hydrogen generation by water splitting using MoS_2 and other transition metal dichalcogenides. *Nano Energy* **2017**, *41*, 49–65. [CrossRef]

10. Eftekhari, A. Electrocatalysts for hydrogen evolution reaction. *Int. J. Hydrog. Eng.* **2017**, *42*, 11053–11077. [CrossRef]

11. Cai, Z.; Liu, B.; Zou, X.; Cheng, H. Chemical vapor deposition growth and applications of two-dimensional materials and their heterostructures. *Chem. Rev.* **2018**, *118*, 6091–6133. [CrossRef]

12. Ye, G.; Gong, Y.; Lin, J.; Li, B.; He, Y.; Pantelides, S.T.; Zhou, W.; Vajtai, R.; Ajayan, P.M. Defects engineered monolayer MoS_2 for improved hydrogen evolution reaction. *Nano Lett.* **2016**, *16*, 1097–1103. [CrossRef]

13. Meng, C.; Lin, M.C.; Du, X.; Zhou, Y. Molybdenum disulfide modified by laser irradiation for catalyzing hydrogen evolution. *ACS Sustain. Chem. Eng.* **2019**, *7*, 6999–7003. [CrossRef]

14. He, Z.; Zhao, R.; Chen, X.; Chen, H.; Zhu, Y.; Su, H.; Huang, S.; Xue, J.; Dai, J.; Cheng, S.; et al. Defect engineering in single-layer MoS_2 using heavy ion irradiation. *ACS Appl. Mater. Interfaces* **2018**, *10*, 42524–42533. [CrossRef]

15. Xia, B.; Wang, T.; Jiang, X.; Zhang, T.; Li, J.; Xiao, W.; Xi, P.; Gao, D.; Xue, D.; Ding, J. Ar^{2+} beam irradiation-induced multivancancies in $MoSe_2$ nanosheet for enhanced electrochemical hydrogen evolution. *ACS Energy Lett.* **2018**, *3*, 2167–2172. [CrossRef]

16. Zhang, P.; Xiang, H.; Tao, L.; Dong, H.; Zhou, Y.; Hu, T.S.; Chen, X.; Liu, S.; Wang, S.; Garaj, S. Chemically activated MoS_2 for efficient hydrogen production. *Nano Energy* **2019**, *57*, 535–541. [CrossRef]

17. Li, G.; Zhang, D.; Qiao, Q.; Yu, Y.; Peterson, D.; Zafar, A.; Kumar, R.; Curtarolo, S.; Hunte, F.; Shannon, S.; et al. All the catalytic active sites of MoS_2 for hydrogen evolution. *J. Am. Chem. Soc.* **2016**, *138*, 16632–16638. [CrossRef]

18. Chang, K.; Hai, X.; Pang, H.; Zhang, H.; Shi, L.; Liu, G.; Liu, H.; Zhao, G.; Li, M.; Ye, J. Targeted synthesis of 2H- and 1T-phase MoS_2 monolayers for catalytic hydrogen evolution. *Adv. Mater.* **2016**, *28*, 10033–10041. [CrossRef]

19. Zhang, J.; Wang, T.; Liu, P.; Liu, Y.; Ma, J.; Gao, D. Enhanced catalytic activities of metal-phase-assisted $1T@2H$-$MoSe_2$ nanosheets for hydrogen evolution. *Electrochim. Acta* **2016**, *217*, 181–186. [CrossRef]

20. Wang, J.; Yan, M.; Zhao, K.; Liao, X.; Wang, P.; Pan, X.; Yang, W.; Mai, L. Field effect enhanced hydrogen evolution reaction of MoS_2 nanosheets. *Adv. Mater.* **2016**, *29*, 1604464. [CrossRef]

21. Guo, J.; Li, F.; Sun, Y.; Zhang, X.; Tang, L. Oxygen-incorporated MoS_2 ultrathin nanosheets grown on graphene for efficient electrochemical hydrogen evolution. *J. Power Sources* **2015**, *291*, 195–200. [CrossRef]

22. Zhu, C.; Xu, Q.; Liu, W.; Ren, Y. CO_2-assisted fabrication of novel heterostructures of h-MoO_3/$1T$-MoS_2 for enhanced photoelectrocatalytic performance. *Appl. Surf. Sci.* **2017**, *425*, 56–62. [CrossRef]

23. Chen, X.; Liu, G.; Zheng, W.; Feng, W.; Cao, W.; Hu, W.; Hu, P. Vertical 2D MoO_2/$MoSe_2$ core–shell nanosheet arrays as high performance electrocatalysts for hydrogen evolution reaction. *Adv. Funct. Mater.* **2016**, *26*, 8537–8544. [CrossRef]

24. Hong, X.; Chan, K.; Tsai, C.; Nørskov, J.K. How doped MoS_2 breaks transition-metal scaling relations for CO_2 electrochemical reduction. *ACS Catal.* **2016**, *6*, 4428–4437. [CrossRef]

25. Chan, K.; Tsai, C.; Hansen, H.A.; Nørskov, J.K. Molybdenum sulfides and selenides as possible electrocatalysts for CO_2 reduction. *ChemCatChem* **2014**, *6*, 1899–1905. [CrossRef]

26. Ji, Y.; Norskov, J.K.; Chan, K. Scaling relations on basal plane vacancies of transition metal dichalcogenides for CO_2 reduction. *J. Phys. Chem. C* **2019**, *123*, 4256–4261. [CrossRef]

27. Jia, P.; Guo, R.; Pan, W.; Huang, C.; Tang, J.; Liu, X.; Qin, H.; Xu, Q. The MoS_2/TiO_2 heterojunction composites with enhanced activity for CO_2 photocatalytic reduction under visible light irradiation. *Colloids Surf. A* **2019**, *570*, 306–316. [CrossRef]

28. Asadi, M.; Kumar, B.; Behranginia, A.; Rosen, B.A.; Baskin, A.; Repnin, N.; Pisasale, D.; Phillips, P.; Zhu, W.; Haasch, R.; et al. Robust carbon dioxide reduction on molybdenum disulphide edges. *Nat. Commun.* **2014**, *5*, 4470. [CrossRef]

29. Asadi, M.; Kim, K.; Liu, C.; Addepalli, A.V.; Abbasi, P.; Yasaei, P.; Phillips, P.; Behranginia, A.; Cerrato, J.M.; Haasch, R.; et al. Nanostructured transition metal dichalcogenide electrocatalysts for CO_2 reduction in ionic liquid. *Science* **2016**, *353*, 467–470. [CrossRef]

30. Bicelli, L.P.; Bozzini, B.; Mele, C.; D'Urzo, L. A review of nanostructural aspects of metal electrodeposition. *Int. J. Electrochem. Sci.* **2008**, *3*, 356–408.

31. Petrova, M.L.; Bojinov, M.; Gadjov, I.H. Electrodeposition of molybdenum oxides from weakly alkaline ammonia-molybdate electrolytes. *Bulg. Chem. Commun.* **2011**, *43*, 60–63.

32. Li, R.; Hao, H.; Huang, T.; Yu, A. Electrodeposited Pd–MoO_x catalysts with enhanced catalytic activity for formic acid electrooxidation. *Electrochim. Acta* **2012**, *76*, 292–299. [CrossRef]

33. Inguanta, R.; Spanò, T.; Piazza, S.; Sunseri, C.; Barreca, F.; Fazio, E.; Neri, F.; Silipigni, L. Electrodeposition and characterization of Mo oxide nanostructures. *Chem. Eng. Trans.* **2015**, *43*, 685–690.

34. Banica, R.; Barvinschi, P.; Vaszilcsin, N.; Nyari, T. A comparative study of the electrochemical deposition of molybdenum oxides thin films on copper and platinum. *J. Alloys Compd.* **2009**, *483*, 402–405. [CrossRef]

35. Soundeswaran, S.; Kumar, O.S.; Babu, S.M.; Ramasamy, P.; Dhanasekaran, R. Influence of ultrasonification in CdS thin film deposition in PCD technique. *Mater. Lett.* **2005**, *59*, 1795–1800. [CrossRef]

36. Podder, J.; Kobayashi, R.; Ichimura, M. Photochemical deposition of Cu_xS thin films from aqueous solutions. *Thin Solid Films* **2005**, *472*, 71–75. [CrossRef]

37. Zhang, W.L.; Zhang, S.; Yang, M.; Chen, T.P. Microstructure of magnetron sputtered amorphous SiO_x films: Formation of amorphous Si core-shell nanoclusters. *J. Phys. Chem. C* **2010**, *114*, 2414–2420. [CrossRef]

38. Wu, S.; Huang, H.; Shang, M.; Du, C.; Wu, Y.; Song, W. High visible light sensitive MoS_2 ultrathin nanosheets for photoelectrochemical biosensing. *Biosens. Bioelectron.* **2017**, *92*, 646–653. [CrossRef]

39. Deng, S.; Zhong, Y.; Zeng, Y.; Wang, Y.; Yao, Z.; Yang, F.; Lin, S.; Wang, X.; Lu, X.; Xia, X.; et al. Directional construction of vertical nitrogen-doped 1T-2H $MoSe_2$/graphene shell/core nanoflake arrays for efficient hydrogen evolution reaction. *Adv. Mater.* **2017**, *29*, 1700748. [CrossRef]

40. Jiang, Q.; Lu, Y.; Huang, Z.; Hu, J. Facile solvent-thermal synthesis of ultrathin $MoSe_2$ nanosheets for hydrogen evolution and organic dyes adsorption. *Appl. Surf. Sci.* **2017**, *402*, 277–285. [CrossRef]

41. Nam, D.; Lee, J.U.; Cheong, H. Excitation energy dependent Raman spectrum of $MoSe_2$. *Sci. Rep.* **2015**, *5*, 17113. [CrossRef]

42. Gu, J.I.; Lee, J.; Rhee, C.K.; Sohn, Y. Enhanced electrochemical hydrogen evolution over defect-induced hybrid MoO_3/$Mo_3O_9 \cdot H_2O$ microrods. *Appl. Surf. Sci.* **2019**, *469*, 348–356. [CrossRef]

43. Xu, X.B.; Sun, Y.; Qiao, W.; Zhang, X.; Chen, X.; Song, X.Y.; Du, Y.W. 3D MoS_2-graphene hybrid aerogels as catalyst for enhanced efficient hydrogen evolution. *Appl. Surf. Sci.* **2017**, *396*, 1520–1527. [CrossRef]

44. Jian, W.; Cheng, X.; Huang, Y.; You, Y.; Zhou, R.; Sun, T.; Xu, J. Arrays of ZnO/MoS_2 nanocables and MoS_2 nanotubes with phase engineering for bifunctional photoelectrochemical and electrochemical water splitting. *Chem. Eng. J.* **2017**, *328*, 474–483. [CrossRef]

45. Zhang, J.; Wu, M.H.; Shi, Z.T.; Jiang, M.; Jian, W.J.; Xiao, Z.; Li, J.; Lee, C.S.; Xu, J. Composition and interface engineering of alloyed $MoS_{2x}Se_{2(1-x)}$ nanotubes for enhanced hydrogen evolution reaction activity. *Small* **2016**, *12*, 4379–4385. [CrossRef]

46. Upadhyay, K.K.; Nguyen, T.; Silva, T.M.; Carmezim, M.J.; Montemor, M.F. Electrodeposited MoO_x films as negative electrode materials for redox supercapacitors. *Electrochim. Acta* **2017**, *225*, 19–28. [CrossRef]

47. Liu, C.; Xie, Z.; Wang, W.; Li, Z.; Zhang, Z. Fabrication of MoO_x film as a conductive anode material for micro-supercapacitors by electrodeposition and annealing. *J. Electrochem. Soc.* **2014**, *161*, A1051–A1057. [CrossRef]

48. Guo, L.; Oskam, G.; Radisic, A.; Hoffmann, P.M.; Searson, P.C. Island growth in electrodeposition. *J. Phys. D* **2011**, *44*, 443001. [CrossRef]

49. Shu, H.; Zhou, D.; Li, F.; Cao, D.; Chen, X. Defect engineering in $MoSe_2$ for the hydrogen evolution reaction: From point defects to edges. *ACS Appl. Mater. Interfaces* **2017**, *9*, 42688–42698. [CrossRef]

50. Ravikumar, C.H.; Nair, G.V.; Muralikrishna, S.; Nagaraju, D.H.; Balakrishna, R.G. Nanoflower like structures of $MoSe_2$ and MoS_2 as efficient catalysts for hydrogen evolution. *Mater. Lett.* **2018**, *220*, 133–135. [CrossRef]

51. Sekizawa, K.; Oh-Ishi, K.; Kataoka, K.; Arai, T.; Suzuki, T.M.; Morikawa, T. Stoichiometric water splitting using p-type Fe_2O_3 based photocathode with the aid of multi-heterojunction. *J. Mater. Chem. A* **2017**, *5*, 6483–6493. [CrossRef]

52. Briggs, D.; Seah, M.P. *Practical Surface Analysis*, 2nd ed.; Wiley and Sons: Hoboken, NJ, USA, 1990; Volume 1.

53. Tosini, G.; Ferguson, I.; Tsubota, K. Effects of blue light on the circadian system and eye physiology. *Mol. Vis.* **2016**, *22*, 61–72. [PubMed]

Article

Band Gap Modulation of Tantalum(V) Perovskite Semiconductors by Anion Control

Young-Il Kim [1,2,*] and Patrick M. Woodward [2]

[1]   Department of Chemistry, Yeungnam University, Gyeongsan 38541, Korea
[2]   Department of Chemistry and Biochemistry, The Ohio State University, Columbus, OH 43210, USA;
      woodward.55@osu.edu
*   Correspondence: yikim@ynu.ac.kr; Tel.: +82-53-810-2353

Received: 11 January 2019; Accepted: 3 February 2019; Published: 7 February 2019

Abstract: Band gap magnitudes and valence band energies of Ta^{5+} containing simple perovskites ($BaTaO_2N$, $SrTaO_2N$, $CaTaO_2N$, $KTaO_3$, $NaTaO_3$, and TaO_2F) were studied by diffuse reflection absorbance measurements, density-functional theoretical calculations, and X-ray photoelectron spectroscopy. As a universal trend, the oxynitrides have wider valence bands and narrower band gaps than isostructural oxides, owing to the N $2p$ contribution to the electronic structure. Visible light-driven water splitting was achieved by using Pt-loaded $CaTaO_2N$, together with a sacrificial agent CH_3OH.

Keywords: perovskite oxynitride; band gap; density-functional theory; water splitting

---

1. Introduction

Transition metal oxynitrides are of interest due to their potential as photocatalysts [1–6], pigments [7,8], battery electrodes [9], high-permittivity dielectrics [10], etc. Such a diverse functionality of oxynitrides is derived largely from the coexistence of O^{2-}/N^{3-} in the anion lattice. As is well established, the conduction and valence bands of simple perovskites AMX_3 are based mostly on the frontier orbitals of M and X, respectively. In the case of oxide perovskites, O $2p$ orbitals participate in the valence band formation near the Fermi level. However, the inclusion of nitrogen, which brings a higher $2p$ orbital energy level than that of oxygen $2p$, can effectively shift the top of the valence band upward resulting in the decreased band gap. It is interesting to note that if the energy difference between O $2p$ (-14.1 eV) and N $2p$ (-11.4 eV) orbitals [11] is reflected onto the valence band edge positions, many of the complex oxynitrides containing Ta^{5+}, Nb^{5+}, or Ti^{4+} would have band gaps falling in the visible light range (3.1~1.8 eV). Such a prospect in optical properties has motivated a number of studies on oxynitride perovskites and related phases, with views to semiconductor developments for visible light-harvesting photocatalysts or non-toxic inorganic pigments [1–8,12–18].

In 2001, Asahi et al. reported the visible light-driven photocatalytic activity of $TiO_{2-x}N_x$ [1], which was followed by a number of studies on oxynitride-type photocatalysts. Promising photocatalysts were identified in various structure types such as simple perovskite ($CaTaO_2N$, $SrTaO_2N$, $BaTaO_2N$, $LaTiO_2N$, $LaTaON_2$, $CaNbO_2N$, $SrNbO_2N$, $BaNbO_2N$), complex perovskites ($LaMg_xTa_{1-x}O_{1+3x}N_{2-3x}$), spinel ($ZnGa_2O_xN_y$), wurtzite ($Ga_{1-x}Zn_xN_{1-x}O_x$), baddeleyite (TaON), and anosovite (Ta_3N_5) [3–6]. Among notable examples, $Ga_{1-x}Zn_xN_{1-x}O_x$ has an absorption edge at ~500 nm and showed a quantum yield of 5.2% for 410 nm light [19]. $BaTaO_2N$-$BaZrO_3$ solid solution could catalyze H_2 evolution from water without sacrificial agents [20]. TaON showed overall water splitting activity with surface modification and appropriate co-catalysts [21]. $LaMg_xTa_{1-x}O_{1+3x}N_{2-3x}$ and $CaTaO_2N$ could achieve overall water splitting with a Rh-Cr mixed oxide co-catalyst [22,23]. However, it was apparent that the photocatalytic behavior of a particular catalyst depended not merely on the composition but the morphology, defects, co-catalysts, and the type of photocatalytic reaction.

In this study, we compare the electronic structures of several Ta^{5+} perovskites having different anion matrices of pure oxide, oxynitride, and oxyfluoride types. The diffuse reflection absorbance spectra for $ATaO_2N$ (A = Ba, Sr, and Ca) are presented along with those of $KTaO_3$, $NaTaO_3$, and TaO_2F, revealing a clear dependence of the semiconductor band gap on the electronegativity of anion components. The density-functional theory (DFT) based computations, combined with the valence level X-ray photoelectron spectroscopy (XPS), confirm that the N $2p$ component plays a critical role in extending the valence band edge in oxynitride compounds. We also present the photocatalytic activity of oxynitride samples tested by examining the water splitting under visible light irradiation.

2. Results and Discussion

Figure 1 displays the diffuse reflection absorbance spectra for simple Ta^{5+} perovskites where the oxynitride phases are found with markedly smaller band gap energies than the others. The optical band gaps were estimated by Shapiro's method [24]. The linear region of the absorption edge was extrapolated to the wavelength axis, where the intersection (zero absorption) was taken as the band gap value. The estimated band gap energies are in the following order: $BaTaO_2N$ (1.8 eV) < $SrTaO_2N$ (2.1 eV) < $CaTaO_2N$ (2.4 eV) < $KTaO_3$ (3.6 eV) < $NaTaO_3$ (4.0 eV) < TaO_2F (4.1 eV).

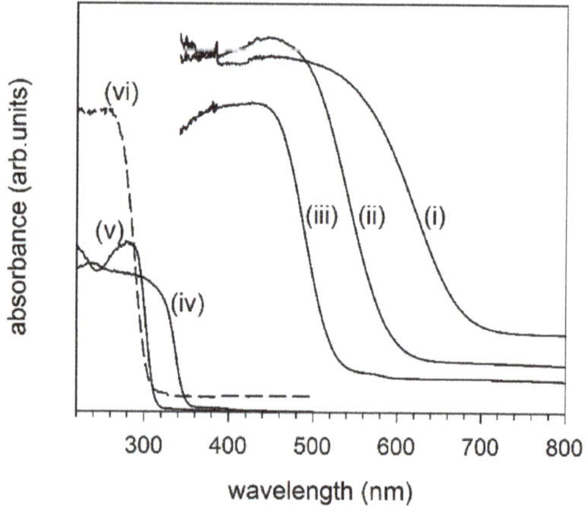

Figure 1. Diffuse reflection absorption spectra for (**i**) $BaTaO_2N$, (**ii**) $SrTaO_2N$, (**iii**) $CaTaO_2N$, (**iv**) $KTaO_3$, (**v**) $NaTaO_3$, and (**vi**) TaO_2F.

For the d^0 perovskites AMX_3, it has been well elucidated that the band gap magnitude depends on (i) electronegativity difference between M cation and X anion, (ii) deviation of $M-X-M$ bond angles away from $180°$, (iii) $M-X$ bond distance, and (iv) electronegativity of A cation [25,26]. The band gap variation among $BaTaO_2N$ (cubic), $SrTaO_2N$ (tetragonal), and $CaTaO_2N$ (orthorhombic) can be explained by the structural distortion factor (ii), in which the more distorted $Ta-(O,N)-Ta$ linkage leads to the narrower band width and the wider band gap. However, the same reasoning cannot be used across distinct anion systems as the cubic $KTaO_3$ has a greater band gap than that of $CaTaO_2N$. In this regard, it can be judged that the control of anion components among N, O, and F, which have well-separated electronegativity values, makes a dominant effect on the resulting electronic structure. The absorbance spectra were also examined by using Tauc plots [27] from which the above oxynitride perovskites were found to be indirect-gap semiconductors.

A detailed aspect of the electronic structural evolution depending on the anion components was studied by band calculations at the DFT level and the XPS measurements. Structural parameters for DFT calculations were taken from the Rietveld refinements for $BaTaO_2N$, $SrTaO_2N$, and $CaTaO_2N$ [10], or from the literature data for $KTaO_3$ [28], $NaTaO_3$ [29], and TaO_2F [30]. Since the computation codes cannot handle mixed occupation of any crystallographic site, ordered O/N (or O/F) distributions were assumed for mixed anion phases.

The density of states (DOS) in $BaTaO_2N$, $SrTaO_2N$, $CaTaO_2N$, $KTaO_3$, and TaO_2F resulted from the calculations using CAmbridge Serial Total Energy Package (CASTEP) and are compared in Figure 2. As previously observed for similar compounds, the computation tends to underestimate the band gap magnitude. Still, it can be well recognized that the width and position of valence bands vary depending on the anion components. Both of the mixed anion systems have widened valence bands due to the $2p$ orbital mixings between O/N or O/F: extended toward a higher energy side for oxynitrides and toward a lower energy side for oxyfluoride. However, the conduction bands of those five compounds were found at fairly similar energy ranges (not shown) since they have the same octahedral cation, Ta. The net result is the effective band gap reduction in oxynitrides, as compared with oxides. On the other hand, for the oxyfluoride derivative, the band gap itself would not change very much to a first approximation.

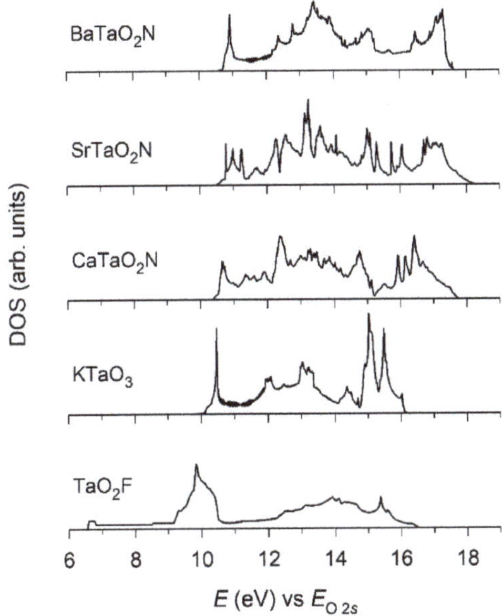

Figure 2. Valence band density of states (DOS) structures for $BaTaO_2N$, $SrTaO_2N$, $CaTaO_2N$, $KTaO_3$, and TaO_2F as calculated using the CAmbridge Serial Total Energy Package (CASTEP) code.

The N $2p$ contribution to band structures of oxynitride compounds can be better viewed by extracting the partial DOS of the component atoms. Figure 3 shows the DOS plots for $BaTaO_2N$ as an example, which was obtained by employing linear muffin-tin orbital (LMTO) calculation. Both O p and N p orbitals were found as the major constituents of the valence band but notably the N character resides primarily at the upper region of the valence band, in agreement with the design concept of these oxynitride perovskites.

Along with the theoretical calculation, an experimental probe was also used to study the valence band structures of $ATaO_2N$ (A = Ba, Sr, Ca), $KTaO_3$, $NaTaO_3$, and TaO_2F. The XPS spectra presented in Figure 4 were collected at near the Fermi level and, therefore, depict the DOS of valence bands. After the energy calibration using the C $1s$ peak energy and background subtraction, the tops of the valence bands were determined as indicated on the plots (Figure 4). The valence band edges of oxynitrides were found to be higher in energy by significant margins ($\approx$1 eV) than the oxides' or oxyfluoride's, which is consistent with the electronic structure calculations. It is, therefore, corroborated both experimentally and theoretically that the hybridization of O 2*p* and N 2*p* orbitals are energetically feasible in the extended solid lattice, and that the partial N/O replacement can be a useful means to reduce the band gap size of oxide semiconductors. Based on the measured band gap magnitudes and the valence band widths, simplified band structures can be proposed for the Ta^{5+} perovskites studied here (Figure 5).

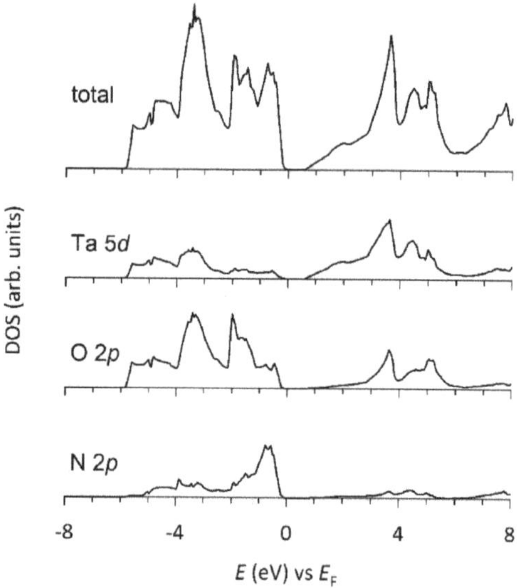

Figure 3. Total and partial DOS for BaTaO$_2$N as calculated using linear muffin-tin orbital (LMTO) code.

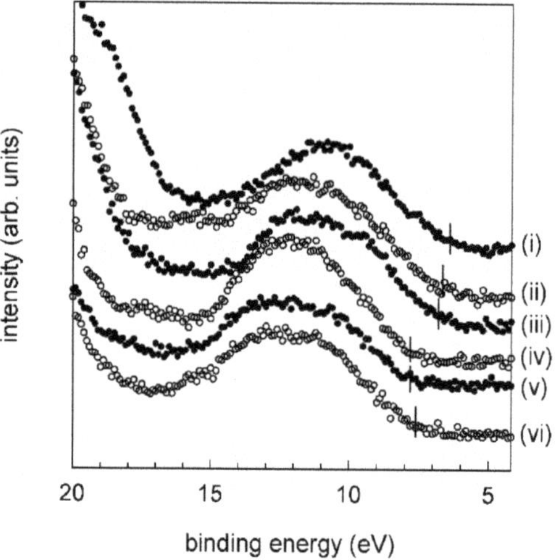

Figure 4. Valence level XPS spectra for (**i**) BaTaO$_2$N, (**ii**) SrTaO$_2$N, (**iii**) CaTaO$_2$N, (**iv**) KTaO$_3$, (**v**) NaTaO$_3$, and (**vi**) TaO$_2$F. Vertical bar on each data indicates the top edge of the valence band.

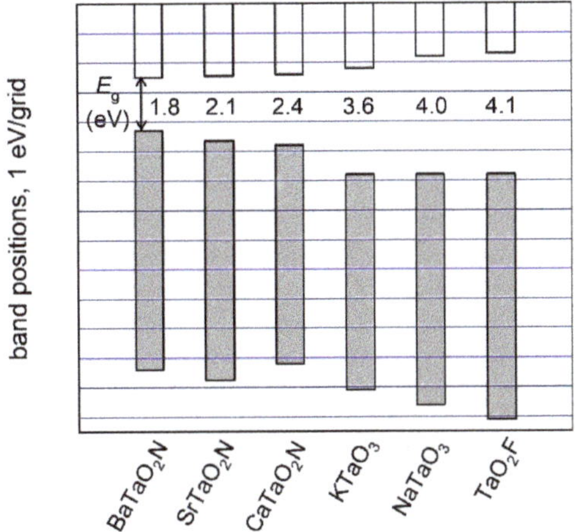

Figure 5. Conduction (unfilled) and valence (shaded) band positions for several simple perovskites with octahedral Ta^{5+}, as deduced from diffuse reflection absorbance and XPS measurements.

The reduced band gaps of oxynitride phases have immediate relevance to the photocatalytic reactivity. In this respect, we tested the water splitting by Pt-loaded oxynitride samples under visible light irradiation. Figure 6 presents the time-dependent H_2 evolution from the Pt-CaTaO$_2$N in H_2O/CH_3OH, along with the result from Pt-TiO$_2$ (P25). Since CH_3OH contains carbon with a formal oxidation number of -2, it can act as a reducing agent that removes O_2 and expedite water decomposition as follows:

$$2H_2O + h\nu \rightarrow 2H_2 + O_2 \tag{1}$$

$$2CH_3OH + O_2 \rightarrow 2CO_2 + 2H_2 \tag{2}$$

The sacrificial agent CH_3OH should boost the generation of H_2 according to the Le Chatelier principle, and also help suppress the $H_2 - O_2$ recombination.

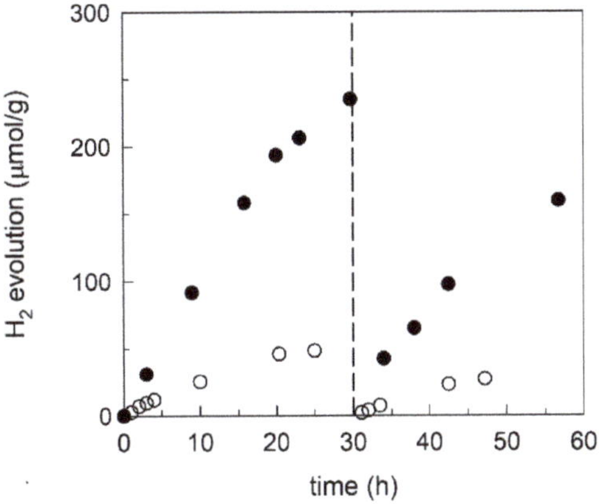

Figure 6. Photocatalytic H_2 evolutions over Pt-loaded powders of CaTaO$_2$N (filled circles) and P25 TiO$_2$ (open circles). At the beginning and after 30 h had elapsed, the reactor vessel was purged with Ar.

As displayed in Figure 6, Pt-CaTaO$_2$N possesses the photocatalytic activity that can be triggered by visible light photons. Using the irradiation source of $\lambda > 395$ nm here, the photocatalytic efficiency of Pt-CaTaO$_2$N is significantly higher than that of Pt-TiO$_2$ (P25), a well-established photocatalyst system. Certainly, the superior performance of Pt-CaTaO$_2$N is attributed to its narrower band gap. As can be found from Figure 1, CaTaO$_2$N can utilize the photons with λ as long as $\approx$500 nm, whereas the absorption by TiO$_2$ (P25) is limited to $\lambda < 400$ nm. The other oxynitride samples Pt-BaTaO$_2$N and Pt-SrTaO$_2$N and an oxide sample Pt-KTaO$_3$ were also examined under the same experimental condition, but none of them produced discernible amounts of H$_2$. The lack of photocatalytic ability in Pt-KTaO$_3$ is simply ascribable to its wide band gap. However, in the cases of BaTaO$_2$N and SrTaO$_2$N, which possess even smaller band gap energies than CaTaO$_2$N, the inferior photocatalytic property can be due to other factors. As one possibility, the valence band edges of BaTaO$_2$N and SrTaO$_2$N might be higher than the O^{2-}/O$_2$ oxidation level, or the H$_2$−O$_2$ recombination might occur so fast as to disallow the observation of water decomposition. Yet, BaTaO$_2$N and SrTaO$_2$N are regarded as promising candidates for visible light photocatalysts that could be well exploited in deliberately designed reaction systems. In the studies by Domen et al., it was demonstrated that the combination of Pt-ATaO$_2$N (A = Ba, Sr, Ca) and Pt-WO$_3$ achieves overall water splitting under visible light in the presence of IO$_3{}^-$/I$^-$ as a shuttle redox mediator [2,12].

3. Materials and Methods

3.1. Sample Syntheses and Crystal Structure

Polycrystalline oxynitride samples ATaO$_2$N (A = Ba, Sr, Ca) were prepared by ammonolysis reaction using BaCO$_3$ (J. T. Baker, 99.8%, Phillipsburg, NJ, USA), SrCO$_3$ (Aldrich, 99.9+%, St. Louis, MO, USA), CaCO$_3$ (Mallinckrodt, 99.95%, Phillipsburg, NJ, USA), and Ta$_2$O$_5$ (Cerac, 99.5%, Milwaukee, WI, USA), as described previously [10]. Quantitative mixture of reagents was heated in anhydrous ammonia (99.99%) at a flow rate of $\approx$50 cm^3/min. Each ammonolytic heating cycle consisted of heating/cooling ramps of 10 °C/min and a dwell step of 20 h at 1000 °C. The heat treatment cycle was repeated 2–5 times to obtain phase pure products. For preparing TaO$_2$F, Ta powder (Alfa Aesar, 99.9%, Karlsruhe, Germany) was dissolved in HF solution (47%) in a Teflon beaker. After evaporating the solvent at 125 °C, white precipitate was washed with distilled water, dried at 150 °C, and finally heated at 500 °C in air for 1 h. Reference compounds KTaO$_3$ (Cerac, 99.9%) and NaTaO$_3$ (Cerac, 99.9%) were used as purchased.

Crystal structure analyses of ATaO$_2$N samples used the synchrotron X-ray powder diffraction patterns collected at the beamline X7A of National Synchrotron Light Source, Brookhaven National Laboratory (Upton, NY, USA). Lattice parameters and atomic coordinates for ATaO$_2$N phases were refined using the Rietveld method as incorporated in the GSAS-GUI software suite [31,32].

3.2. Electronic Structure and Photocatalytic Property

Diffuse reflectance data were recorded and converted to absorbance using a spectrophotometer (Perkin Elmer, Lambda 20, Waltham, MA, USA) equipped with a 50-mm Labsphere integrating sphere over the spectral range 200–900 nm. The band gap energies were determined from Shapiro's method [24] of extrapolating the onset of absorption to the wavelength axis.

DFT-based computations were performed using the CASTEP program as embodied in Accelrys Materials Studio [33]. Norm-conserving nonlocal pseudo-potentials were generated using the Kerker scheme with a kinetic energy cutoff of 400 eV. A convergence criterion of 0.02 meV was applied for the energy change per atom. Electron exchange and correlation were described using the Perdew-Wang generalized gradient approximation (PW91-GGA) [34]. For BaTaO$_2$N, the total and partial densities of states were also calculated using a computation code, Stuttgart LMTO version 47, developed by Anderson and co-workers [35,36]. The program employs a TB-LMTO-ASA (tight binding linear muffin-tin orbital atomic sphere approximation) algorithm. Integrations over k space were performed using the tetrahedron method with a total of 40 irreducible k points from a $6 \times 6 \times 6$ grid of reducible k points.

Valence band structures of $ATaO_2N$ (A = Ba, Sr, Ca), $KTaO_3$, $NaTaO_3$, and TaO_2F were experimentally studied by XPS at near Fermi energy level, using a V. G. Scientific spectrometer equipped with a Mg K_α source (1253.6 eV) and operated at 9 kV and 20 mA with a base pressure of $\approx 2 \times 10^{-9}$ Torr. Shirley method [37] was used for the data smoothening and background removal from the raw XPS spectra.

Photocatalytic activity of $CaTaO_2N$, in comparison with that of TiO_2 (Degussa P25) [38], was examined for the water decomposition using visible light irradiation. To focus on the photocatalytic H_2 evolution, Pt was employed as a co-catalyst [39]. For preparing the Pt-impregnated catalyst, sample powder was stirred in an aqueous solution of $H_2PtCl_6 \cdot 6H_2O$ ($[Pt^{4+}] \approx 0.4$ mM) under ultraviolet (UV) irradiation for 24 h, rinsed, and dried at room temperature. Thus, the obtained Pt-loaded catalyst (≈ 50 mg) was suspended in a mixture of 35 mL H_2O and 0.6 mL MeOH contained in a 43.5 mL quartz vessel, which was sealed with a latex septum and filled with ≈ 1 atm of Ar. The photocatalytic reaction was induced by external illumination with an Oriel Xe lamp (24 V, 7 A) through a liquid filter and a long-pass filter (λ_{cutoff} = 395 nm), and was monitored using a gas chromatograph (Shimadzu, GC-14A, Tokyo, Japan) with Ar (99.998%) carrier gas. By using the liquid filter with a circulating water cooler, the reaction vessel was kept from the heating effect of infrared light component.

4. Conclusions

It was shown, using six simple perovskites with octahedral Ta^{5+}, that the semiconductor band gap can be widely modulated by the electronegativity of anion components. The band gap generally widens from oxynitrides to oxides to oxyfluorides, and in most cases, the d^0 oxynitride phases have band gaps corresponding to visible light energy. Band structure calculations by the DFT method and XPS measurements indicate that the N $2p$ component contributes to extend the top of the valence band in oxynitrides, making a principal distinction from the oxides' electronic structures. The photocatalytic H_2 generation from H_2O was observed by using Pt-$CaTaO_2N$ and a sacrificial electron donor CH_3OH under visible light.

Author Contributions: P.M.W. designed the research and reviewed the draft. Y.-I.K. prepared and characterized the samples and wrote the draft.

Funding: This research was funded by the National Research Foundation of Korea (NRF-2015R1D1A1A01056591) through the Basic Science Research Program.

Conflicts of Interest: The authors declare no conflict of interest.

References

1. Asahi, R.; Morikawa, T.; Ohwaki, T.; Aoki, K.; Taga, Y. Visible-Light Photocatalysis in Nitrogen-Doped Titanium Oxides. *Science* **2001**, *293*, 269–271. [CrossRef] [PubMed]

2. Higashi, M.; Abe, R.; Takata, T.; Domen, K. Photocatalytic Overall Water Splitting under Visible Light Using $ATaO_2N$ (A = Ca, Sr, Ba) and WO_3 in a IO_3^-/I^- Shuttle Redox Mediated System. *Chem. Mater.* **2009**, *21*, 1543–1549. [CrossRef]

3. Takata, T.; Pan, C.; Domen, K. Recent progress in oxynitride photocatalysts for visible-light-driven water splitting. *Sci. Technol. Adv. Mater.* **2015**, *16*, 033506. [CrossRef] [PubMed]

4. Wu, Y.; Lazic, P.; Hautier, G.; Persson, K.; Ceder, G. First principles high throughput screening of oxynitrides for water-splitting photocatalysts. *Energy Environ. Sci.* **2013**, *6*, 157–168. [CrossRef]

5. Zhang, P.; Zhang, J.; Gong, J. Tantalum-based semiconductors for solar water splitting. *Chem. Soc. Rev.* **2014**, *43*, 4395–4422. [CrossRef] [PubMed]

6. Ahmed, M.; Xinxin, G. A review of metal oxynitrides for photocatalysis. *Inorg. Chem. Front.* **2016**, *3*, 578–590. [CrossRef]

7. Jansen, M.; Letschert, H.P. Inorganic yellow-red pigments without toxic metals. *Nature* **2000**, *404*, 980–982. [CrossRef]

8. Tessier, F.; Marchand, R. Ternary and higher order rare-earth nitride materials: Synthesis and characterization of ionic-covalent oxynitride powders. *J. Solid State Chem.* **2003**, *171*, 143–151. [CrossRef]

9. Cabana, J.; Dupre, N.; Gillot, F.; Chadwick, A.V.; Grey, C.P.; Palacin, M.R. Synthesis, Short-Range Structure, and Electrochemical Properties of New Phases in the Li$-$Mn$-$N$-$O System. *Inorg. Chem.* **2009**, *48*, 5141–5153. [CrossRef]

10. Kim, Y.-I.; Woodward, P.M.; Baba-Kishi, K.Z.; Tai, C.W. Characterization of the Structural, Optical, and Dielectric Properties of Oxynitride Perovskites AMO_2N (A = Ba, Sr, Ca; M = Ta, Nb). *Chem. Mater.* **2004**, *16*, 1267–1276. [CrossRef]

11. Harrison, W.A. *Electronic Structure and the Properties of Solids*; W. H. Freeman and Company: San Francisco, CA, USA, 1980; p. 50.

12. Higashi, M.; Abe, R.; Teramura, K.; Takata, T.; Ohtani, B.; Domen, K. Two step water splitting into H_2 and O_2 under visible light by $ATaO_2N$ (A = Ca, Sr, Ba) and WO_3 with $IO_3{}^-/I^-$ shuttle redox mediator. *Chem. Phys. Lett.* **2008**, *452*, 120–123. [CrossRef]

13. Yashima, M.; Yamada, H.; Maeda, K.; Domen, K. Experimental visualization of covalent bonds and structrural disorder in a gallium zince oxynitride photocatalyst $(Ga_{1-x}Zn_x)(N_{1-x}O_x)$: Origin of visible light absorption. *Chem. Commun.* **2010**, *46*, 2379–2382. [CrossRef] [PubMed]

14. Hisatomi, T.; Katayama, C.; Moriya, Y.; Minegishi, T.; Katayama, M.; Nishiyama, H.; Yamada, T.; Domen, K. Photocatalytic oxygen evolution using $BaNbO_2N$ modified with cobalt oxide under photoexcitation up to 740 nm. *Energy Environ. Sci.* **2013**, *6*, 3595–3599. [CrossRef]

15. Xiao, M.; Wang, S.; Thaweesak, S.; Luo, B.; Wang, L. Tantalum (Oxy)Nitride: Narrow Bandgap Photocatalysts for Solar Hydrogen generation. *Engineering* **2017**, *3*, 365–378. [CrossRef]

16. Mukherji, A.; Seger, B.; Lu, G.Q.; Wang, L. Nitrogen Doped $Sr2Ta2O7$ Coupled with Graphene Sheets as Photocatalysts for Increased Photocatalytic Hydrogen Production. *ACS Nano* **2011**, *5*, 3483–3492. [CrossRef] [PubMed]

17. Chen, S.; Yang, J.; Ding, C.; Li, R.; Jin, S.; Wang, D.; Han, H.; Zhang, F.; Li, C. Nitrogen-doped layered oxide $Sr_5Ta_4O_{15-x}N_x$ for water reduction and oxidation under visible light irradiation. *J. Mater. Chem. A* **2013**, *1*, 5651–5659. [CrossRef]

18. Wang, X.; Maeda, K.; Thomas, A.; Takanabe, K.; Xin, G.; Carlsson, J.M.; Domen, K.; Antonietti, M. A metal-free polymeric photocatalyst for hydrogen production from water under visible light. *Nat. Mater.* **2009**, *8*, 76–80. [CrossRef] [PubMed]

19. Kubota, J.; Domen, K. Photocatalytic Water Splitting Using Oxynitride and Nitride Semiconductor Powders for Production of Solar Hydrogen. *Electrochem. Soc. Interface* **2013**, *22*, 57–62. [CrossRef]

20. Matoba, T.; Maeda, K.; Domen, K. Activation of $BaTaO_2N$ photocatalyst for enhanced non-sacrificial hydrogen evolution from water under visible light by forming a solid solution with $BaZrO_3$. *Chem. Eur. J.* **2011**, *17*, 14731–14735. [CrossRef] [PubMed]

21. Maeda, K.; Lu, D.; Domen, K. Direct water splitting into hydrogen and oxygen under visible light by using modified TaON photocatalysts with d^0 electronic configuration. *Chem. Eur. J.* **2013**, *19*, 4986–4991. [CrossRef] [PubMed]

22. Pan, C.; Takata, T.; Nakabayashi, M.; Matsumoto, T.; Shibata, N.; Ikuhara, Y.; Domen, K. A complex perovskite-type oxynitride: The first photocatalyst for water splitting operable at up to 600 nm. *Angew. Chem. Int. Ed.* **2015**, *54*, 2955–2959. [CrossRef] [PubMed]

23. Xu, J.; Pan, C.; Takata, T.; Domen, K. Photocatalytic overall water splitting on the perovskite-type transition metal oxynitride $CaTaO_2N$ under visible light irradiation. *Chem. Commun.* **2015**, *51*, 7191–7194. [CrossRef]

24. Shapiro, I.P. Determination of the forbidden zone width from diffuse reflection spectra. *Opt. Specktroskopiya* **1958**, *4*, 256–260.

25. Cox, P.A. *The Electronic Structure and Chemistry of Solids*; Oxford University Press: Oxford, UK, 1987; pp. 45–72.

26. Eng, H.W.; Barnes, P.W.; Auer, B.M.; Woodward, P.M. Investigations of the electronic structure of d0 transition metal oxides belonging to the perovskite family. *J. Solid State Chem.* **2003**, *175*, 94–109. [CrossRef]

27. Tauc, J. Optical properties and electronic structure of amorphous Ge and Si. *Mater. Res. Bull.* **1968**, *3*, 37–46. [CrossRef]

28. Zhurova, E.A.; Ivanov, Y.; Zavodnik, V.; Tsirelson, V. Electron density and atomic displacements in $KTaO_3$. *Acta Cryst. B* **2000**, *56*, 594–600. [CrossRef]

29. Kennedy, B.J.; Prodjosantoso, A.K.; Howard, C.J. Powder neutron diffraction study of the high temperature phase transitions in $NaTaO_3$. *J. Phys. Condens. Matter* **1999**, *11*, 6319–6327. [CrossRef]

30. Frevel, L.K.; Rinn, H.W. The crystal structure of NbO_2F and TaO_2F. *Acta Cryst.* **1956**, *9*, 626–627. [CrossRef]

31. Larson, A.C.; von Dreele, R.B. *General Structure Analysis System (GSAS)—Report LAUR 86-748*; Los Alamos National Laboratory: Los Alamos, NM, USA, 1994.
32. Toby, B. *EXPGUI*, a graphical user interface for *GSAS*. *J. Appl. Cryst.* **2001**, *34*, 210–213. [CrossRef]
33. Segall, M.D.; Lindan, P.J.D.; Probert, M.J.; Pickard, C.J.; Hasnip, P.J.; Clark, S.J.; Payne, M.C. First-principles simulation: Ideas, illustrations and the CASTEP code. *J. Phys. Condens. Matter* **2002**, *14*, 2717–2744. [CrossRef]
34. Perdew, J.P.; Chevary, J.A.; Vosko, S.H.; Jackson, K.A.; Pederson, M.R.; Singh, D.J.; Fiolhais, C. Atoms, molecules, solids, and surfaces: Applications of the generalized gradient approximation for exchange and correlation. *Phys. Rev. B* **1992**, *46*, 6671–6687. [CrossRef]
35. Anderson, O.K.; Jepsen, O. Explicit, First-Principles Tight-Binding Theory. *Phys. Rev. Lett.* **1984**, *53*, 2571–2574. [CrossRef]
36. Anderson, O.K.; Pawlowska, Z.; Jepsen, O. Illustration of the linear-muffin-tin-orbital tight-binding representation: Compact orbitals and charge density in Si. *Phys. Rev. B* **1986**, *34*, 5253–5269. [CrossRef]
37. Shirley, D.A. High-Resolution X-Ray Photoemission Spectrum of the Valence Bands of Gold. *Phys. Rev. B* **1972**, *5*, 4709–4714. [CrossRef]
38. Ohtani, B.; Prieto-Mahaney, O.O.; Li, D.; Abe, R. What is Degussa (Evonik) P25? Crystalline composition analysis, reconstruction from isolated pure particles and photocatalytic activity test. *J. Photochem. Photobiol. A Chem.* **2010**, *216*, 179–182. [CrossRef]
39. Yang, J.Y.; Wang, D.; Han, H.; Li, C. Roles of cocatalysts in photocatalysis and photoelectrocatalysis. *Acc. Chem. Res.* **2013**, *46*, 1900–1909. [CrossRef] [PubMed]

Review

Photocatalytic Hydrogen Evolution via Water Splitting: A Short Review

Yifan Zhang [1], Young-Jung Heo [1], Ji-Won Lee [1], Jong-Hoon Lee [1], Johny Bajgai [2], Kyu-Jae Lee [2,*] and Soo-Jin Park [1,*]

[1] Department of Chemistry and Chemical Engineering, Inha University, 100 Inharo, Incheon 22212, Korea; zyf9106@gmail.com (Y.Z.); heoyj1211@gmail.com (Y.-J.H.); lj529@naver.com (J.-W.L.); boy834@naver.com (J.-H.L.)

[2] Department of Environmental Medical Biology, Wonju College of Medicine, Yonsei University, Wonju 26426, Korea; johnybajgai@gmail.com

* Correspondence: medbio9@gmail.com (K.-J.L.); sjpark@inha.ac.kr (S.-J.P.); Tel.: +82-32-860-7234 (S.-J.P.); Fax: +82-32-860-5604 (S.-J.P.)

Received: 23 October 2018; Accepted: 8 December 2018; Published: 12 December 2018

Abstract: Photocatalytic H_2 generation via water splitting is increasingly gaining attention as a viable alternative for improving the performance of H_2 production for solar energy conversion. Many methods were developed to enhance photocatalyst efficiency, primarily by modifying its morphology, crystallization, and electrical properties. Here, we summarize recent achievements in the synthesis and application of various photocatalysts. The rational design of novel photocatalysts was achieved using various strategies, and the applications of novel materials for H_2 production are displayed herein. Meanwhile, the challenges and prospects for the future development of H_2-producing photocatalysts are also summarized.

Keywords: photocatalysis; H_2 generation; water splitting; solar energy

1. Introduction

The development of renewable green energy sources is a critical challenge for modern society. H_2 is environmentally friendly, renewable, and considered to be an ideal candidate for an economically and socially sustainable fuel [1–6], and was previously regarded as an alternative energy source. Interestingly, some researchers also found that H_2-rich water has neuron effects owing to its antioxidant properties. Although the deep mechanism is not clear, more and more researchers made an effort to study the biological function of H_2 [7–21]. To date, almost all H_2 gas production processes in the industry are based on natural gas, coal, petroleum, or water electrolysis. These traditional preparation methods are limited due to the associated CO_2 emissions and high energy consumption. Hence, it is urgent to develop a low-cost method for efficient H_2 generation and, thus, support the emerging H_2 economy.

The sun provides an energy output of ~3×10^{24} J per year, which is approximately 12,000 times higher than the current energy demand. Therefore, solar energy can act as a sustainable alternative energy source in the future. To date, the transformation of solar energy into H_2 via water splitting is deemed as a desirable H_2 preparation method to solve the energy crisis [22,23].

The proper use of H_2 requires insight into the physical properties of H_2 molecules. As we know, the lengths and strengths of hydrogen bonds are exquisitely sensitive to temperature and pressure. Meanwhile, the charges of H_2 molecules also vary with temperature [24] because the spin direction of the nucleus in the H_2 molecule changes depending on the temperature, and an energy difference occurs between H_2 molecules. The *para*-H_2 fraction changes with temperature, and it is necessary to understand the characteristics of H_2 molecules according to temperature [25]. During the reaction, hydrogen can be used safely at room temperature; however, it is rather dangerous in high-temperature environments.

As we know, H_2 gas, often called dihydrogen or molecular H_2, is a highly flammable gas with a wide range of concentrations between 4% and 75% by volume. Meanwhile, H_2 is the world's lightest gas. The density of H_2 is only 1/14 of that of air. At 0 °C, the density of H_2 is only 0.0899 g/L at standard atmospheric pressure, which is the smallest-molecular-weight substance; it is mainly used as a reducing agent. The enthalpy of combustion is about −286 kJ/mol, which can be displayed by the following equation: $2H_2(g) + O_2(g) \rightarrow 2H_2O(l) + 572$ kJ (286 kJ/mol). Currently, H_2 is the main industrial raw material and the most important industrial gas. It has various applications in the petrochemical, electronic, and metallurgical industry, as well as in food processing, float glass, fine organic synthesis, aerospace, and other fields. At the same time, H_2 is also an ideal secondary energy source. Owing to the properties of H_2, the aerospace industry uses liquid H_2 as fuel. Now, it is common to produce H_2 from water gas rather than using high-energy-consuming water. The produced H_2 is used in large quantities in the cracking reaction of the petrochemical industry and the production of ammonia. Unfortunately, all H_2 production methods are highly energy (thermal and electrical) demanding, which limits their application. Thus, it is crucial to find a new method of H_2 production.

Fujishima and Honda first reported photocatalytic water splitting using a TiO_2 electrode in 1972 [26]. Research on solar H_2 production attracted researchers in various fields, such as (1) chemists for the design and synthesis of various catalysts to investigate structure–property relationships; (2) physicists to fabricate semiconductor photocatalysts with novel electronic structures, as predicted by theoretical calculation; and (3) material scientists to construct unique photocatalytic materials with novel structures and morphologies [27–30]. When photocatalysts are illuminated at wavelengths which are suitable to their band gap energy, after the excitation, the charge carriers will either combine or transfer to the surface of the photocatalysts to participate in photocatalytic reactions. For the generation of efficient semiconductor photocatalysts, long-lived charge carriers and high stability are required [31–33].

Significant developments were made toward H_2 generation via water splitting over the last several decades by a number of talented researchers [34–38].

Herein, we attempt to sum up the advances achieved to date. Therefore, we briefly summarize the background related to various photocatalysts for H_2 generation and the achievements of high-efficiency photocatalysts. The main synthesis routes and modifications for adjusting the band structure to harvest light and enhance charge separation are also discussed.

2. Principle of H_2 Generation via Water Splitting

In the pioneering study by Fujishima and Honda [27], electrochemical cells were made up for the splitting of the water into H_2 and O_2, as shown in Figure 1. While the TiO_2 electrode was under ultraviolet (UV) light irradiation, water oxidation (oxygen evolution) occurred on its surface, while the reduction reaction (H_2 evolution) occurred on the surface platinum black electrode. With this study in mind, semiconductor photocatalysts were later developed by Bard et al. in their design of a novel photocatalytic system.

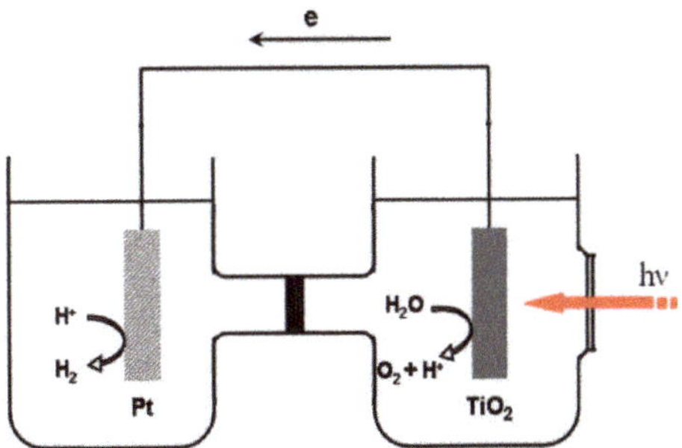

Figure 1. Schematic of a photoelectrochemical cell (PEC). Reproduced with permission from Reference [26]; copyright (1972), Nature Publishing Group.

Figure 2a shows a display of hydrogen evolution by photocatalysts. The photocatalytic reaction occurring on the semiconductor photocatalysts can be divided into three parts: (1) obtaining photons with energy exceeding that of the photocatalyst's band gap, generating electron and hole pairs; (2) separating carriers by migration in the semiconductor photocatalyst; and (3) reaction between these carriers and H_2O [39–46]. In addition, electron–hole pairs will combine with each other simultaneously. As shown in Figure 2b, while photocatalysts are involved in hydrogen evolution, the lowest position of the conduction band (CB) should be lower than the reduction position of H_2O/H_2, while the position of the valence band (VB) should be higher than the potential of H_2O/O_2 [47–50].

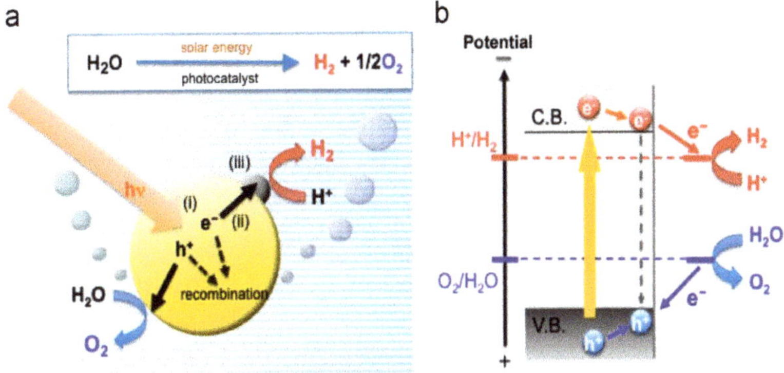

Figure 2. Schematic illustration of hydrogen evolution over photocatalysts. Reproduced with permission from Reference [39]; copyright (2014), Elsevier.

Various photocatalysts were reported to decompose water into H_2 and O_2 (Equation (1)). As we know, the hydrogen evolution reaction can be separated into two parts: oxidation for the evolution of O_2 (Equation (2)) and water reduction to produce H_2 (Equation (3)) [51–56]:

$$H_2O \rightarrow 2H_2 + O_2 \qquad\qquad \Delta E^0 = 1.23\ V \tag{1}$$

$$H_2O \rightarrow 4H^+ + 4e^- + O_2 \qquad\qquad E^0 = +1.23\ V\ vs.\ NHE,\ pH = 0 \tag{2}$$

$$4H^+ + 4e^- \rightarrow 2H_2 \qquad\qquad \Delta E^0 = 0\ V\ vs.\ NHE,\ pH = 0 \tag{3}$$

3. Photocatalysts for Water Splitting

Many photocatalysts were created as photocatalysts for hydrogen evolution. Based on these species, they can be divided into three major parts: (1) graphene-based photocatalyst; (2)

graphitic carbon nitride (g-C3N4)-based photocatalysts; and (3) heterojunction photocatalysts (semiconductor–semiconductor or semiconductor–(metal, element)).

3.1. Graphene-Based Photocatalysts

Recently, graphene-based photocatalysts attracted significant attention for enhancing photocatalytic H_2 production performance. Graphene is used to enhance photocatalytic efficiency owing to its novel structure and electrochemical properties (Figure 3).

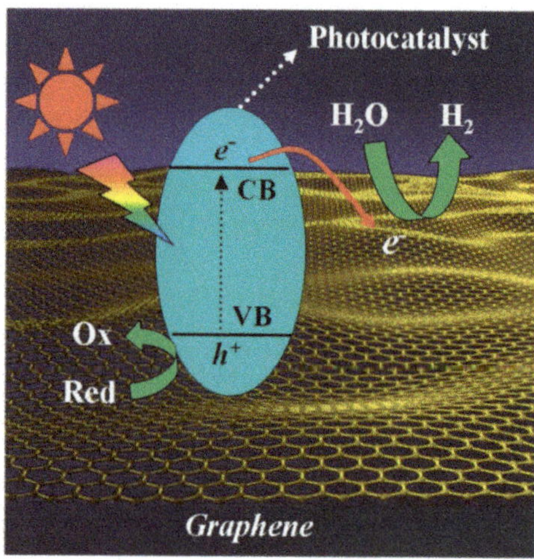

Figure 3. Proposed mechanism of graphene-based photocatalysts. Reproduced with permission from Reference [56]; copyright (2013), American Chemical Society.

To date, many reports regarding the synthesis of graphene-based photocatalysts with improved photocatalytic efficiency were published. Graphene is a well-known two-dimensional (2D) material, which can improve surface area, and its 2D membrane-like structure imparts unique electrochemical properties [57–60]. Generally speaking, photocatalysts prepared by simple physical mixing with graphene will involve only a bit of direct contact with the graphene sheets. This small amount of contact between the photocatalyst and graphene results in weak interactions and inhibits charge transfer rates. Hence, the synthesis of photocatalysts with more interactions is highly needed.

Previously, Kim et al. synthesized novel graphene oxide (GO)-TiO_2 photocatalysts [58] in 2013, comprising a core–shell nanostructure with enhanced photocatalytic efficiency (Figure 4). The improved H_2 production activity compared to that of TiO_2 revealed that the utilization of the core–shell structure enhanced photocatalytic efficiency. This novel structural design offers three-dimensional (3D) close contact between the materials and provides more active sites, which will enhance the charge separation rate and H_2 production efficiency [61–63].

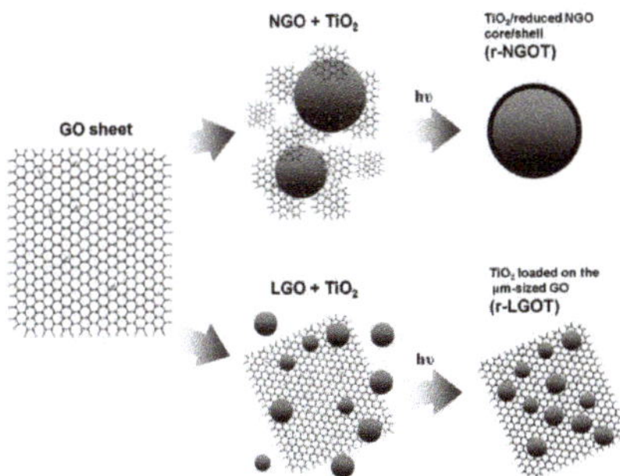

Figure 4. Schematic display of synthetic process of graphene oxide (GO)/TiO$_2$ and TiO$_2$/GO. Reproduced with permission from Reference [57]; copyright (2012), American Chemical Society.

Currently, many researchers are more interested in visible-light-driven photocatalysts, which are achieved using band-gap modification or taking graphene as a photosensitizer to broaden the visible-light adsorption range [64–66]. Significant efforts were conducted for building visible-light response systems because of the UV-only response of TiO$_2$, and its nontoxic properties [67]. Recently, it was found that graphene regulating TiO$_2$ involves visible-light adsorption activity. The carbon-layered structure of graphene with enriched π electrons forms bonds with titanium atoms. As a result, this strong interaction will shift the band position and reduce the band gap [68–70]. Lee et al. [71] also achieved a lower band gap using a graphene/TiO$_2$ photocatalyst. The improved photocatalytic efficiency of the graphene/TiO$_2$ composite owes to the band-gap regulation, which consequently promotes charge transfer rates through the graphene sheets.

3.2. g-C$_3$N$_4$-Based Photocatalysts

Currently, carbon-nitride-based photocatalysts receive significant attention for their photocatalytic H$_2$ generation owing to a unique electronic structure (Figure 5) [72–77]. This section summarizes recent significant achievements in building C$_3$N$_4$-based photocatalysts for H$_2$ evolution. Methods including nanostructure regulation, band-gap modification, dye sensitization, and heterojunction fabrication are highlighted herein.

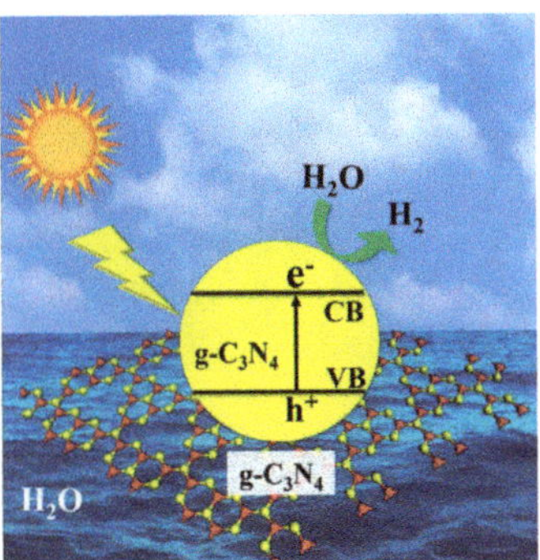

Figure 5. Proposed mechanism of graphitic carbon nitride (g-C$_3$N$_4$)-based photocatalysts. Reproduced with permission from Reference [72]; copyright (2014), American Chemical Society.

Recently, carbon nitride attracted significant attention following the pioneering research of Wang et al. in 2009 for photocatalytic hydrogen evolution [78,79]. The assumed structure of C$_3$N$_4$ is a 2D framework with the tri-*s*-triazine linked by tertiary amines (Figure 6); it is thermally stable and chemically stable. Pioneering studies regarded g-C$_3$N$_4$ as a visible-light-driven phorocatalyst with a band gap of approximately 2.7 eV and an appropriate band position for water splitting [80–85]. Hence, g-C_3N_4 is an ideal candidate for photocatalytic H_2 evolution.

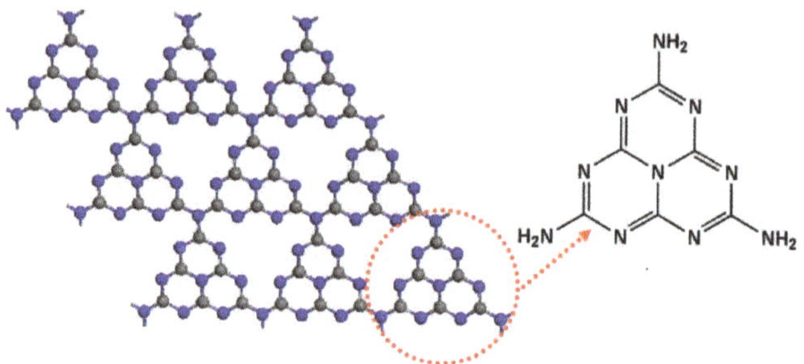

Figure 6. Schematic display of the structure of g-C_3N_4. Reproduced with permission from Reference [72]; copyright (2014), American Chemical Society.

H_2 generation performance using g-C_3N_4 can be promoted with noble-metal particles such as Au or Pd, which obtain electrons in the CB to inhibit the charge recombination rate [86–90]. Many researchers are developing metal-free photocatalysts for H_2 evolution, and recent reports involved the introduction of non-noble-metal catalysts into g-C_3N_4 photocatalysts, displaying enhanced photocatalytic performance compared to noble-metal catalysts [91–95]. Hou et al. [86] synthesized MoS_2/g-C_3N_4 composite photocatalysts (Figure 7) in 2018. MoS_2/g-C_3N_4 increased the surface area and decreased the barrier when the electrons transported, thereby improving the charge transfer rate. The formation of band alignment enabled electron transfer from the CB (g-C_3N_4) to MoS_2. Therefore, the MoS_2/gC_3N_4 nanojunction significantly enhanced H_2 evolution efficiency, achieving the highest H_2 evolution rate and an optimum quantum efficiency of up to 2.1% (420 nm), which was higher than g-C_3N_4/Pt.

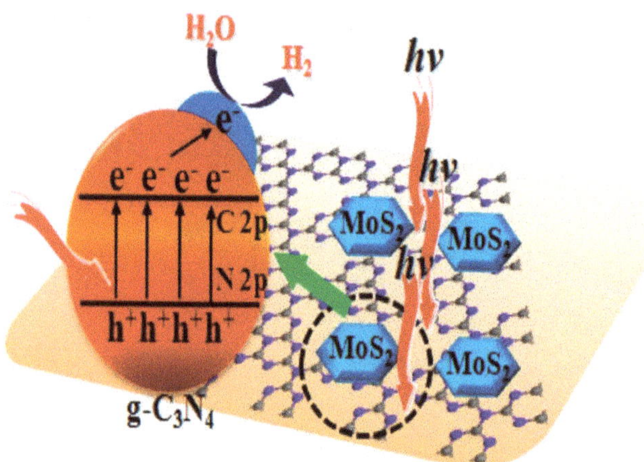

Figure 7. Schematic display of charge transfer on MoS_2/g-C_3N_4 heterostructures during water splitting. Reproduced with permission from Reference [72]; copyright (2014), American Chemical Society.

3.3. Metal-Loading-Based Photocatalysts

Metal loading is also regarded as a useful method for photocatalytic enhancement. Song et al. [96] constructed Ag-rGO-TiO_2 composite photocatalysts (Figure 8) in 2018. In order to analyze the photocatalytic mechanism of the architectural Ag-TiO_2 and Ag-rGO-TiO_2 composites, their structures with Ag nanocubes for light absorption and TiO_2 nanosheets were well displayed. The difference between Ag-TiO$_2$ and Ag-rGO-TiO$_2$ is the interface between Ag nanocubes and TiO$_2$ nanosheets, which enhances the electron transfer capability. For Ag-TiO$_2$, the direct contact between the two materials results in the formation of Ag (100)/(001) TiO$_2$ interface. Meanwhile, for Ag-rGO-TiO$_2$, both Ag(100)/rGO and rGO/(001) TiO$_2$ interfaces are formed by rGO. As mentioned above, the synergistic effect of Ag(100)/rGO and rGO/(001)TiO$_2$ interfaces, rather than the Ag(100)/(001) TiO$_2$ interface, offers quicker electron transfer. As shown in Figure 8, no Schottky barrier is formed between Ag and TiO$_2$, and the hot electrons on the surface of TiO$_2$ flow back to Ag and then recombine with holes. Meanwhile, for the Ag-rGO-TiO$_2$ sample, no barrier is necessary to facilitate the electron transfer. The electrons generated on the surface of Ag nanocubes with smaller work function flow to rGO via a contact so as to equilibrate the electron Fermi distribution on the interface [97,98]. Moreover, the rGO nanosheets can act as conductive channels, further transferring the electron to the rGO/TiO$_2$ interface. Owing to the light absorption of rGO, the transferred electrons within the rGO nanosheets can be further transferred to the CB of TiO$_2$ under light excitation. The proposed photocatalytic mechanism of Ag-rGO-TiO$_2$ is illustrated in Figure 8.

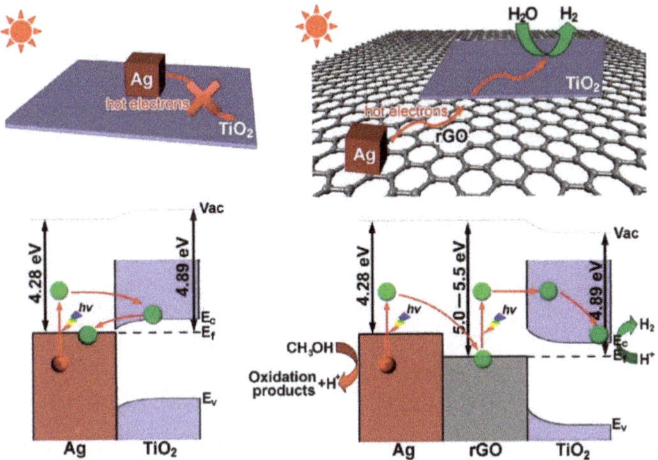

Figure 8. Schematic illustrating photocatalytic mechanism for Ag-TiO$_2$ and Ag-rGO-TiO$_2$ samples under visible-light irradiation. Reproduced with permission from Reference [95]; copyright (2018), Elsevier.

3.4. Z-Scheme Photocatalysts

An illustration of Z-scheme water splitting is shown in Figure 9. During an H$_2$ evolution reaction, the reactions which happen on the surface of photocatalysts include the reduction of protons by CB electrons and the oxidation of an electron donor (D) by VB holes, yielding the corresponding electron acceptor (A), as follows:

$$2H^+ + 2e^- \rightarrow H_2 \; (photoreduction \; of \; H^+ \; to \; H_2)$$

$$D + nh^+ \rightarrow A \; (photooxidation \; of \; D \; to \; A)$$

On the other hand, the forward reactions on an O$_2$ evolution photocatalyst are as follows:

$$A + ne^- \rightarrow D \; (photoreduction \; of \; A \; to \; D)$$

$$2H_2O + 4h^+ \rightarrow O_2 + 4H^+ \; ((photooxidation \; of \; H_2O \; to \; O_2)$$

where the electron acceptor generated by the paired H_2 evolution photocatalyst is converted to D, and the water oxidation process occurs via the valence band holes. Thus, the water-splitting process can be achieved.

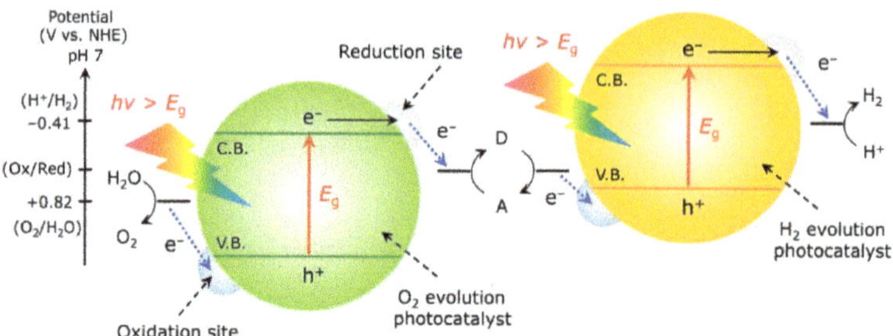

Figure 9. Diagram of photocatalytic water splitting using a Z-Scheme system. Reproduced with permission from Reference [98]; Copyright (2010), American Chemical Society.

Amal et al. reported a Z-scheme photocatalytic water-splitting system using $Ru/SrTiO_3$ and partially reduced GO (PRGO)/$BiVO_4$ (Figure 10) in 2011 [100]. As described in the report, the PRGO/$BiVO_4$ (O_2 photocatalyst) and $Ru/SrTiO_3$:Rh (H_2 photocatalyst) were attached due to surface charge modification in acidic conditions, as depicted in Figure 10. Under irradiation, electrons are excited from the VB ($BiVO_4$) or an impurity level in Rh ($Ru/SrTiO_3$:Rh) to the CB. We can indicate that the PRGO does not contribute to the electron and hole generation. In other words, the RGO in this work acts as an electron conductor. PRGO transfers the electrons from the CB of $BiVO_4$ to the $Ru/SrTiO_3$:Rh. Meanwhile, the electrons in $Ru/SrTiO_3$:Rh reduce the water to H_2 on the surface of the Ru co-catalyst, while the holes left in $BiVO_4$ oxidize the water to O_2. Additionally, the PRGO provides a pathway for photogenerated electrons in the $BiVO_4$ photocatalyst. Each reaction can migrate as follows: reduction of water, transfer of electrons to PRGO, and transfer of holes to PRGO for oxidation. Because the majority of the photocatalyst surface is surrounded by water and only relatively small portions are in contact with PRGO [101], most electrons in $Ru/SrTiO_3$:Rh and holes in $BiVO_4$ are used for water splitting.

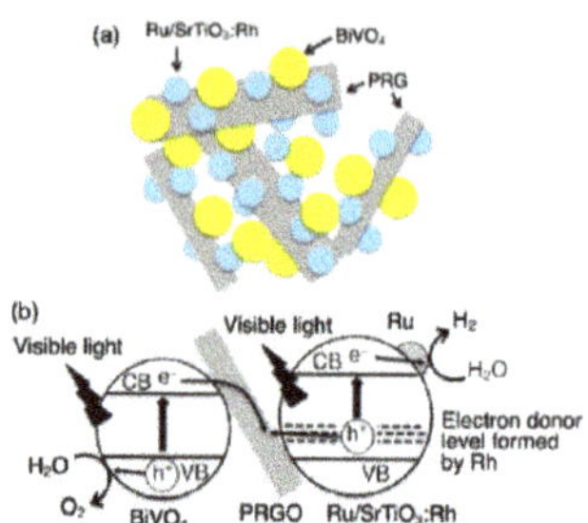

Figure 10. (a) Schematic display of a suspension of $Ru/SrTiO_3$ and partially reduced GO (PRGO)/$BiVO_4$ in water. (b) Mechanism of water splitting using Z-scheme system consisting of $Ru/SrTiO_3$ and PRGO/$BiVO_4$ under irradiation. Reproduced with permission from Reference [99]; copyright (2018), American Chemical Society.

3.5. Defect Engineering Photocatalyst

Among the various photocatalyst designs, the defect engineering strategy is regarded as an important way of modifying the photocatalysts. Defects are places where the atoms or molecules in the materials are disrupted, and they greatly influence photocatalytic performance. The defects in the lattice of photocatalysts not only act as an electron–hole recombination center, but also break the electronic structure and display a scattering center for electron and hole travel. Nevertheless, the positive effect of defects in photocatalytic performance enhancement were also recognized with the development of defect photocatalysts and the development of the photocatalytic field.

Chen et al. reported the synthesis of a bismuth subcarbonate ($Bi_2O_2CO_3$, BOC) with controllable defect density (BOC-X) (Figure 11) in 2018. The BOC-X with defect density displayed a photocatalytic nitrogen fixation of 957 μmol$\cdot$L^{-1} under irradiation within 4 h, which was 9.4 times higher than that of pristine BOC. This photocatalytic performance enhancement of BOC-X can be attributed to the surface defects. These defects contribute to the defect levels in the forbidden band, which improves the light harvest percentage. Meanwhile, surface defects can also inhibit the electron–hole recombination rate to promote the separation efficiency of charge carriers. Photocatalytic nitrogen fixation by BOC-X is displayed in Figure 11. If the light energy is higher than the band-gap energy, the electrons on the VB surface of BOC-X are transferred to the CB and react with N_2 to form NH_3. Moreover, some of the VB electrons are transferred to the defect level and then react with N_2. However, if the light energy is lower than the band-gap energy, the electrons of BOC-X are also excited from VB to the defect level and then participate in the reaction. Defects modulate the band gap of BOC-X and improve the light absorption range, thereby enhancing the carrier transport, and leading to photocatalytic enhancement.

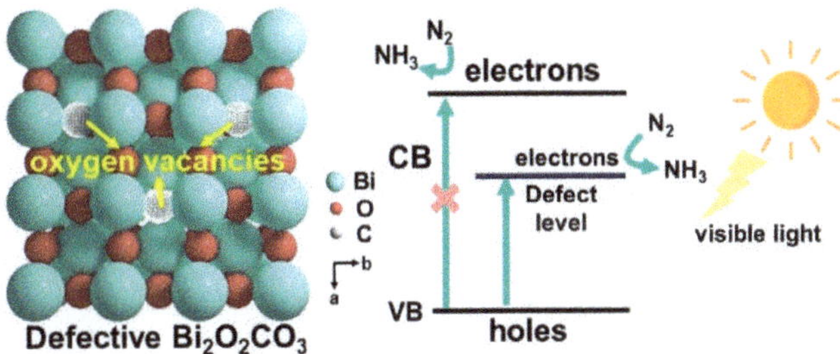

Figure 11. Mechanism of photocatalytic nitrogen fixation on defective $Bi_2O_2CO_3$. Reproduced with permission from Reference [101]; copyright (2010), American Chemical Society.

3.6. Heterojunction Photocatalysts

During the H_2 evolution reaction, the formed electron–hole charges are transferred to the surface of the photocatalyst for the next step of the reaction or recombine with each other [102–106]. To better reveal this point, we assumed it as a simple case [107]: the influence of gravity on a man jumping (Figure 12a,b). When a man (electron) jumps from the ground (VB) to the sky (CB), it can return to the floor immediately (recombination of the electron and hole) owing to gravity. In order to let the people rise off the floor (separation of the charge carrier pairs), an instrument (semiconductor B) can be used (Figure 12c,d). Subsequently, the previously mentioned people can drop to the instrument rather than the ground (inhibition of the electron and hole pair recombination). Although the inhibition of electron–hole recombination rate is an urgent issue, it can be achieved via suitable construction of materials. Many methods were conducted to achieve better electron–hole pair separation rate, such as element combining [108,109], metal doping [110,111], or the use of heterojunctions [112,113]. Among these strategies, heterojunctions were proven to be the most desirable method for achieving efficient photocatalysis due to their improved separation ability of electron–hole pairs (Figure 12d).

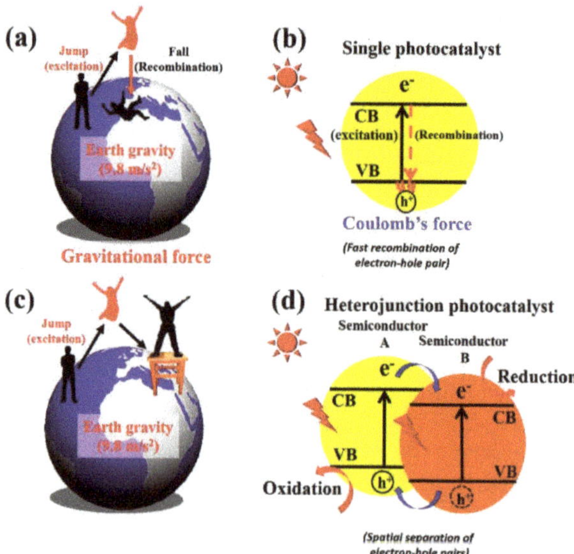

Figure 12. Schematic display of (**a**) the influence of gravity on a person jumping, (**b**) electron–hole pair combination using a photocatalyst, (**c**) utilization of a stool to keep the person from returning to the ground, (**d**) electron–hole pairs separated in composite catalyst. Reproduced with permission from Reference [105]; copyright (2010), John Wiley & Sons, Inc.

A heterojunction is regarded as the connection between two kinds of photocatalysts with different band structures, which leads to a new band arrangement [114,115]. Generally, three kinds of composite photocatalysts are developed (Figure 13). As shown in Figure 13a, the CB and VB of A are a bit over and under the band position of B, respectively [116]. As a result, when the light irradiates, the generated electrons and holes are transferred to the CB and VB of B. Because the generated electrons and holes move to the same photocatalyst, the recombination rate of electron–hole pairs is not efficiently inhibited. The photocatalytic process happens on photocatalyst B with a mild potential requirement; thus, the photocatalytic ability of the photocatalyst using this heterojunction will be lower than others. As dispalyed in Figure 13b, the band positions of CB and VB are over that of photocatalyst B. Hence, during the photocatalytic reaction, the generated electron moves to photocatalyst B, while the holes are transferred to photocatalyst A, which leads to the formation of long lived electron–hole pairs [117–119]. Parallel to Figure 13a, the photocatalytic performance of the type-II composite photocatalysts is inhibited by the redox process occurring on B. Meanwhile, as displayed in Figure 13c, the band structure of type-III composite photocatalysts is parallel to type II, apart from the interlaced gap changing into non-overlapping band gaps [120,121]. Thus, the generated electron–hole pairs cannot be transferred between the two photocatalysts, resulting in them being inappropriate for long lived electron–hole pair separation. We can determine that the type-II heterojunctions are desirable for enhancing redox ability due to their optimum structure for long-lived electron–hole separation. In previous reports, great efforts were conducted to synthesize type-II composite photocatalysts, including g-C_3N_4/TiO_2 [122], WO_3/$BiVO_4$ [123], WO_3/g-C_3N_4 [124], and $BiPO_4$/g-C_3N_4 [125].

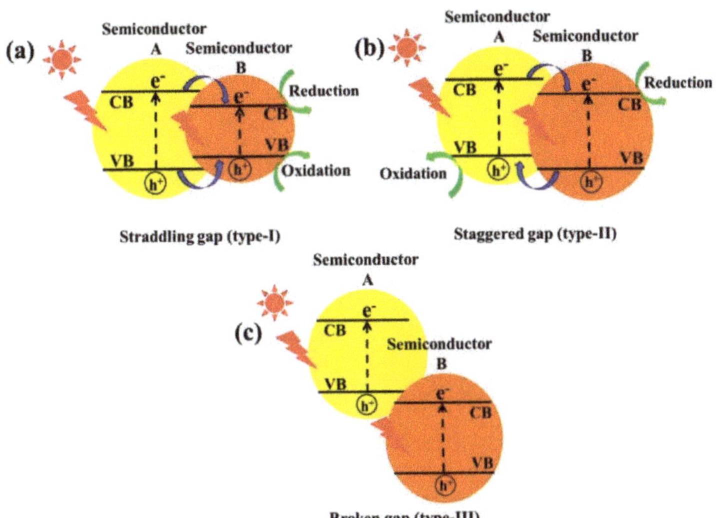

Figure 13. Schematic display of three kinds of electron–hole pair separation among composite photocatalysts: (**a**) type-I, (**b**) type-II, (**c**) type-III heterojunctions. Reproduced with permission from Reference [105]; copyright (2017), John Wiley & Sons, Inc.

Yu et al. designed CdS/NiS composites photocatalysts using various heterojunctions in 2012, which greatly enhanced the hydrogen evolution performance. As shown in Figure 14a, around 20 nm of NiS particles were loaded onto the CdS uniformly, which supported a close connection between CdS and NiS. The formation of p–n heterojunctions facilitates charge transfer between the NiS and CdS, and inhibits charge-carrier recombination (Figure 14b,c). We can see that the holes left on the n-type catalyst are transferred to the p-type catalyst, providing a negative specie. The electron–hole pair distribution keeps moving until a Fermi-level equilibrium is achieved [126–128]. The generated active species move through the internal electric field of the composite photocatalysts, resulting in long-lived electron–hole pair separation rates. Thus, the electron–hole recombination rate is efficiently inhibited owing to the synergistic effect between the two photocatalysts. The photocatalytic H_2 production rate over CdS/NiS composite photocatalysts with 5 wt.% NiS was found to be higher than that of the CdS and 1 wt.% Pt/CdS (Figure 14d). More NiS doping resulted in a reduction in photocatalytic efficiency due to NiS catalysts reducing the number of redox sites during the reaction.

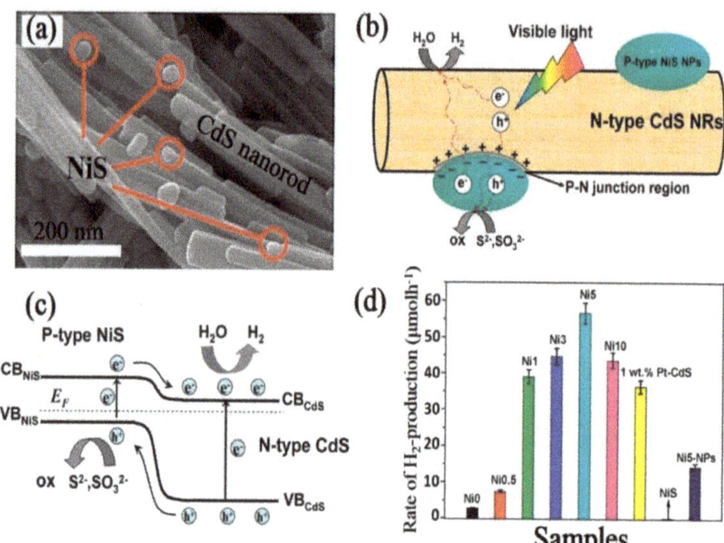

Figure 14. (**a**) SEM image of CdS/NiS composite catalysts; (**b**,**c**) illustration of electron–hole pairs with CdS/NiS composite photocatalysts; (**d**) contrast of photocatalytic efficiency of CdS with different NiS content. Reproduced with permission from Reference [105]; copyright (2017), John Wiley & Sons, Inc.

4. Summary and Perspectives

Over the last several decades, photocatalysis was shown to be a promising method for H_2 production. Even though the principles controlling photocatalytic activity in the developed semiconductors were identified, several aspects remain unclear. Therefore, practical applications and the commercialization of photocatalytic H_2 production require further research. Meanwhile, the charge transfer among photocatalysts due to the influence of structure and electrochemical properties is also not very clear, while the influence of various preparation methods on the catalytic performance is not well understood. The development of improved photocatalysts will benefit from advances in science. Improved building of novel co-catalysts will arise from using efficient catalysts. Many researches are underway investigating new synthesis methods for sample preparation and novel system construction. Herein, we concluded the most prominent achievements associated with H_2 production via photocatalysis. We hope this report will assist further research efforts regarding the development of photocatalysts.

Funding: This research was supported by the Korea Evaluation Institute of Industrial Technology (KEIT) through the Carbon Cluster Construction project [10083586, Development of petroleum-based graphite fibers with ultra-high thermal conductivity] funded by the Ministry of Trade, Industry, & Energy (MOTIE, Korea), and the Commercialization Promotion Agency for R&D Outcomes (COMPA) funded by the Ministry of Science and ICT (MSIT) [2018_RND_002_0064, Development of 800 mA·h·g^{-1} pitch carbon coating materials].

Conflicts of Interest: The authors declare no conflict of interest.

References

1. Park, S.J.; Lee, S.Y. A study on hydrogen-storage behaviors of nickel-loaded mesoporous MCM-41. *J. Colloid Interface Sci.* **2010**, *346*, 194–198. [CrossRef] [PubMed]
2. Catapan, R.C.; Cancino, L.R.; Oliveira, A.A.M.; Schwarz, C.O.; Nitschke, H.; Frank, T. Potential for onboard hydrogen production in an direct injection ethanol fueled spark ignition engine with EGR. *Fuel* **2018**, *234*, 441–446. [CrossRef]

3. Im, J.S.; Park, S.-J.; Kim, T.; Lee, Y.-S. Hydrogen storage evaluation based on investigations of the catalyticproperties of metal/metal oxides in electrospun carbon fibers. *Int. J. Hydrogen Energy* **2009**, *34*, 3382–3388. [CrossRef]

4. Park, S.-J.; Lee, S.-Y. Hydrogen storage behaviors of platinum-supported multi-walled carbon nanotubes. *Int. J. Hydrogen Energy* **2010**, *35*, 13048–13054. [CrossRef]

5. Lee, S.-Y.; Park, S.-J. Effect of platinum doping of activated carbon on hydrogen storage behaviors of metal-organic frameworks-5. *Int. J. Hydrogen Energy* **2011**, *36*, 8381–8387. [CrossRef]

6. Baykara, S.Z. Hydrogen: A brief overview on its sources, production and environmental impact. *Int. J. Hydrogen Energy* **2018**, *43*, 10605–10614. [CrossRef]

7. Chen, M.; Cui, W.; Zhu, K.; Xie, Y.; Zhang, C.; Shen, W. Hydrogen-rich water alleviates aluminum-induced inhibition of root elongation in alfalfa via decreasing nitric oxide production. *J. Hazard. Mater.* **2014**, *267*, 40–47. [CrossRef] [PubMed]

8. Cui, W.; Gao, C.; Fang, P.; Lin, G.; Shen, W. Alleviation of cadmium toxicity in Medicago sativa by hydrogen-rich water. *J. Hazard. Mater.* **2013**, *260*, 715–724. [CrossRef] [PubMed]

9. Gao, Q.; Song, H.; Wang, X.T.; Liang, Y.; Xi, Y.J.; Gao, Y.; Guo, Q.J.; LeBaron, T.; Luo, Y.X.; Li, S.C.; et al. Molecular hydrogen increases resilience to stress in mice. *Sci. Rep.* **2017**, *7*, 9625. [CrossRef]

10. Giuliani, D.; Ottani, A.; Zaffe, D.; Galantucci, M.; Strinati, F.; Lodi, R.; Guarini, S. Hydrogen sulfide slows down progression of experimental Alzheimer's disease by targeting multiple pathophysiological mechanisms. *Neurobiol. Learn. Mem.* **2013**, *104*, 82–91. [CrossRef]

11. Huang, C.S.; Kawamura, T.; Toyoda, Y.; Nakao, A. Recent advances in hydrogen research as a therapeutic medical gas. *Free Radic. Res.* **2010**, *44*, 971–982. [CrossRef] [PubMed]

12. Kumar, R.; Kumar, A.; Langstrom, B.; Darreh-Shori, T. Discovery of novel choline acetyltransferase inhibitors using structure-based virtual screening. *Sci. Rep.* **2017**, *7*, 16287. [CrossRef]

13. Li, J.; Wang, C.; Zhang, J.H.; Cai, J.M.; Cao, Y.P.; Sun, X.J. Hydrogen-rich saline improves memory function in a rat model of amyloid-beta-induced Alzheimer's disease by reduction of oxidative stress. *Brain Res.* **2010**, *1328*, 152–161. [CrossRef]

14. Luo, Q.; Lin, Y.X.; Yang, P.P.; Wang, Y.; Qi, G.B.; Qiao, Z.Y.; Li, B.N.; Zhang, K.; Zhang, J.P.; Wang, L.; et al. A self-destructive nanosweeper that captures and clears amyloid beta-peptides. *Nat. Commun.* **2018**, *9*, 1802. [CrossRef] [PubMed]

15. Nagata, K.; Nakashima-Kamimura, N.; Mikami, T.; Ohsawa, I.; Ohta, S. Consumption of molecular hydrogen prevents the stress-induced impairments in hippocampus-dependent learning tasks during chronic physical restraint in mice. *Neuropsychopharmacol. Off. Publ. Am. Coll. Neuropsychopharmacol.* **2009**, *34*, 501–508. [CrossRef]

16. Nakayama, M.; Itami, N.; Suzuki, H.; Hamada, H.; Yamamoto, R.; Tsunoda, K.; Osaka, N.; Nakano, H.; Maruyama, Y.; Kabayama, S.; et al. Novel haemodialysis (HD) treatment employing molecular hydrogen (H2)-enriched dialysis solution improves prognosis of chronic dialysis patients: A prospective observational study. *Sci. Rep.* **2018**, *8*, 254. [CrossRef] [PubMed]

17. Nishimaki, K.; Asada, T.; Ohsawa, I.; Nakajima, E.; Ikejima, C.; Yokota, T.; Kamimura, N.; Ohta, S. Effects of Molecular Hydrogen Assessed by an Animal Model and a Randomized Clinical Study on Mild Cognitive Impairment. *Curr. Alzheimer Res.* **2018**, *15*, 482–492. [CrossRef] [PubMed]

18. Ohsawa, I.; Ishikawa, M.; Takahashi, K.; Watanabe, M.; Nishimaki, K.; Yamagata, K.; Katsura, K.; Katayama, Y.; Asoh, S.; Ohta, S. Hydrogen acts as a therapeutic antioxidant by selectively reducing cytotoxic oxygen radicals. *Nat. Med.* **2007**, *13*, 688–694. [CrossRef]

19. Xiao, H.W.; Li, Y.; Luo, D.; Dong, J.L.; Zhou, L.X.; Zhao, S.Y.; Zheng, Q.S.; Wang, H.C.; Cui, M.; Fan, S.J. Hydrogen-water ameliorates radiation-induced gastrointestinal toxicity via MyD88's effects on the gut microbiota. *Exp. Mol. Med.* **2018**, *50*, e433. [CrossRef]

20. Zhang, L.M.; Jiang, C.X.; Liu, D.W. Hydrogen sulfide attenuates neuronal injury induced by vascular dementia via inhibiting apoptosis in rats. *Neurochem. Res.* **2009**, *34*, 1984–1992. [CrossRef] [PubMed]

21. Zhang, Y.; Su, W.J.; Chen, Y.; Wu, T.Y.; Gong, H.; Shen, X.L.; Wang, Y.X.; Sun, X.J.; Jiang, C.L. Effects of hydrogen-rich water on depressive-like behavior in mice. *Sci. Rep.* **2016**, *6*, 23742. [CrossRef] [PubMed]

22. Liu, S.; Xin, Z.-J.; Lei, Y.-J.; Yang, Y.; Yan, X.-Y.; Lu, Y.-B.; Li, C.-B.; Wang, H.-Y. Thin Copper-Based Film for Efficient Electrochemical Hydrogen Production from Neutral Aqueous Solutions. *ACS Sustain. Chem. Eng.* **2017**, *5*, 7496–7501. [CrossRef]

23. Wang, Z.; Yang, X.; Yang, T.; Zhao, Y.; Wang, F.; Chen, Y.; Zeng, J.H.; Yan, C.; Huang, F.; Jiang, J.-X. Dibenzothiophene Dioxide Based Conjugated Microporous Polymers for Visible-Light-Driven Hydrogen Production. *ACS Catal.* **2018**, *8*, 8590–8596. [CrossRef]

24. Zhu, K.; Kang, S.-Z.; Qin, L.; Han, S.; Li, G.; Li, X. Novel and Highly Active Potassium Niobate-Based Photocatalyst for Dramatically Enhanced Hydrogen Production. *J. Am. Chem. Soc.* **2005**, *127*, 11447–11453. [CrossRef]

25. Hibino, T.; Kobayashi, K.; Ito, M.; Ma, Q.; Nagao, M.; Fukui, M.; Teranishi, S. Kinetics of the Interconversion of Parahydrogen and Orthohydrogen Catalyzed by Paramagnetic Complex Ions. *J. Am. Chem. Soc.* **2005**, *127*, 11447–11453.

26. Fujishima, A.; Honda, K. Electrochemical Photolysis of Water at a Semiconductor Electrode. *Nature* **1972**, *238*, 37–38. [CrossRef] [PubMed]

27. Zhang, P.; Song, T.; Wang, T.; Zeng, H. In-situ synthesis of Cu nanoparticles hybridized with carbon quantum dots as a broad spectrum photocatalyst for improvement of photocatalytic H_2 evolution. *Appl. Catal. B Environ.* **2017**, *206*, 328–335. [CrossRef]

28. Zhang, P.; Song, T.; Wang, T.; Zeng, H. Plasmonic Cu nanoparticle on reduced graphene oxide nanosheet support: An efficient photocatalyst for improvement of near-infrared photocatalytic H_2 evolution. *Appl. Catal. B Environ.* **2018**, *225*, 172–179. [CrossRef]

29. Zhang, P.; Wang, T.; Zeng, H. Design of Cu-Cu_2O/g-C_3N_4 nanocomponent photocatalysts for hydrogen evolution under visible light irradiation using water-soluble Erythrosin B dye sensitization. *Appl. Surf. Sci.* **2017**, *391*, 404–414. [CrossRef]

30. Zhang, P.; Song, T.; Wang, T.; Zeng, H. Effectively extending visible light absorption with a broad spectrum sensitizer for improving the H_2 evolution of in-situ Cu/g-C_3N_4 nanocomponents. *Int. J. Hydrogen Energy* **2017**, *42*, 14511–14521. [CrossRef]

31. Zhang, Y.; Park, M.; Kim, H.Y.; Ding, B.; Park, S.J. A facile ultrasonic-assisted fabrication of nitrogen-doped carbon dots/BiOBr up-conversion nanocomposites for visible light photocatalytic enhancements. *Sci. Rep.* **2017**, *7*, 45086. [CrossRef] [PubMed]

32. Zhang, Y.; Park, S.-J. Bimetallic AuPd alloy nanoparticles deposited on MoO_3 nanowires for enhanced visible-light driven trichloroethylene degradation. *J. Catal.* **2018**, *361*, 238–247. [CrossRef]

33. Zhang, Y.; Park, S.-J. Au–pd bimetallic alloy nanoparticle-decorated $BiPO_4$ nanorods for enhanced photocatalytic oxidation of trichloroethylene. *J. Catal.* **2017**, *355*, 1–10. [CrossRef]

34. Yu, H.; Xue, Y.; Hui, L.; Zhang, C.; Li, Y.; Zuo, Z.; Zhao, Y.; Li, Z.; Li, Y. Efficient Hydrogen Production on a 3D Flexible Heterojunction Material. *Adv. Mater.* **2018**, *30*, e1707082. [CrossRef]

35. Zhang, P.; Song, T.; Wang, T.; Zeng, H. Fabrication of a non-semiconductor photocatalytic system using dendrite-like plasmonic CuNi bimetal combined with a reduced graphene oxide nanosheet for near-infrared photocatalytic H_2 evolution. *J. Mater. Chem. A* **2017**, *5*, 22772–22781. [CrossRef]

36. Lin, L.; Ren, W.; Wang, C.; Asiri, A.M.; Zhang, J.; Wang, X. Crystalline carbon nitride semiconductors prepared at different temperatures for photocatalytic hydrogen production. *Appl. Catal. B Environ.* **2018**, *231*, 234–241. [CrossRef]

37. Im, J.S.; Kwon, O.; Kim, Y.H.; Park, S.-J.; Lee, Y.-S. The effect of embedded vanadium catalyst on activated electrospun CFs for hydrogen storage. *Microporous Mesoporous Mater.* **2008**, *115*, 514–521. [CrossRef]

38. Yi, H.; Huang, D.; Qin, L.; Zeng, G.; Lai, C.; Cheng, M.; Ye, S.; Song, B.; Ren, X.; Guo, X. Selective prepared carbon nanomaterials for advanced photocatalytic application in environmental pollutant treatment and hydrogen production. *Appl. Catal. B Environ.* **2018**, *239*, 408–424. [CrossRef]

39. Ismail, A.A.; Bahnemann, D.W. Photochemical splitting of water for hydrogen production by photocatalysis: A review. *Sol. Energy Mater. Sol. Cells* **2014**, *128*, 85–101. [CrossRef]

40. Cai, J.; Shen, J.; Zhang, X.; Ng, Y.H.; Huang, J.; Guo, W.; Lin, C.; Lai, Y. Light-Driven Sustainable Hydrogen Production Utilizing TiO_2 Nanostructures: A Review. *Small Methods* **2018**, 1800184. [CrossRef]

41. Ventura-Espinosa, D.; Sabater, S.; Carretero-Cerdán, A.; Baya, M.; Mata, J.A. High Production of Hydrogen on Demand from Silanes Catalyzed by Iridium Complexes as a Versatile Hydrogen Storage System. *ACS Catal.* **2018**, *8*, 2558–2566. [CrossRef]

42. Ji, L.; Lv, C.; Chen, Z.; Huang, Z.; Zhang, C. Nickel-Based (Photo) Electrocatalysts for Hydrogen Production. *Adv. Mater.* **2018**, *30*, e1705653. [CrossRef] [PubMed]

43. Ma, Y.; Dong, X.; Wang, Y.; Xia, Y. Decoupling Hydrogen and Oxygen Production in Acidic Water Electrolysis Using a Polytriphenylamine-Based Battery Electrode. *Angewandte Chem.* **2018**, *57*, 2904–2908. [CrossRef] [PubMed]

44. Wang, B.; Zeng, C.; Chu, K.H.; Wu, D.; Yip, H.Y.; Ye, L.; Wong, P.K. Enhanced Biological Hydrogen Production from Escherichia coli with Surface Precipitated Cadmium Sulfide Nanoparticles. *Adv. Energy Mater.* **2017**, *7*, 1700611. [CrossRef]

45. Xue, Z.; Shen, Y.; Li, P.; Zhang, Y.; Li, J.; Qin, B.; Zhang, J.; Zeng, Y.; Zhu, S. Key Role of Lanthanum Oxychloride: Promotional Effects of Lanthanum in NiLaOy/NaCl for Hydrogen Production from Ethyl Acetate and Water. *Small* **2018**, *14*, e1800927. [CrossRef] [PubMed]

46. Zhang, Y.; Yang, H.M.; Park, S.-J. Synthesis and characterization of nitrogen-doped TiO_2 coatings on reduced graphene oxide for enhancing the visible light photocatalytic activity. *Curr. Appl. Phys.* **2018**, *18*, 163–169. [CrossRef]

47. Zhang, Y.; Park, M.; Kim, H.-Y.; Park, S.-J. In-situ synthesis of graphene oxide/BiOCl heterostructured nanofibers for visible-light photocatalytic investigation. *J. Alloy Compd.* **2016**, *686*, 106–114. [CrossRef]

48. Huang, J.; Li, G.; Zhou, Z.; Jiang, Y.; Hu, Q.; Xue, C.; Guo, W. Efficient photocatalytic hydrogen production over Rh and Nb codoped TiO_2 nanorods. *Chem. Eng. J.* **2018**, *337*, 282–289. [CrossRef]

49. Kim, W.; Monllor-Satoca, D.; Chae, W.-S.; Mahadik, M.A.; Jang, J.S. Enhanced photoelectrochemical and hydrogen production activity of aligned CdS nanowire with anisotropic transport properties. *Appl. Surf. Sci.* **2019**, *463*, 339–347. [CrossRef]

50. Han, J.; Liu, Y.; Dai, F.; Zhao, R.; Wang, L. Fabrication of CdSe/$CaTiO_3$ nanocomposties in aqueous solution for improved photocatalytic hydrogen production. *Appl. Surf. Sci.* **2018**, *459*, 520–526. [CrossRef]

51. Zhang, Y.; Park, M.; Kim, H.Y.; Ding, B.; Park, S.-J. In-situ synthesis of nanofibers with various ratios of BiOClx/BiOBry/BiOIz for effective trichloroethylene photocatalytic degradation. *Appl. Surf. Sci.* **2016**, *384*, 192–199. [CrossRef]

52. Pipitone, G.; Tosches, D.; Bensaid, S.; Galia, A.; Pirone, R. Valorization of alginate for the production of hydrogen via catalytic aqueous phase reforming. *Catal. Today* **2018**, *304*, 153–164. [CrossRef]

53. Hibino, T.; Kobayashi, K.; Ito, M.; Nagao, M.; Fukui, M.; Teranishi, S. Direct electrolysis of waste newspaper for sustainable hydrogen production: An oxygen-functionalized porous carbon anode. *Appl. Catal. B Environ.* **2018**, *231*, 191–199. [CrossRef]

54. Park, S.; Kim, B.; Lee, Y.; Cho, M. Influence of copper electroplating on high pressure hydrogen-storage behaviors of activated carbon fibers. *Int. J. Hydrogen Energy* **2008**, *33*, 1706–1710. [CrossRef]

55. Im, J.S.; Park, S.-J.; Lee, Y.-S. Superior prospect of chemically activated electrospun carbon fibers for hydrogen storage. *Mater. Res. Bull.* **2009**, *44*, 1871–1878. [CrossRef]

56. Xiang, Q.; Yu, J. Graphene-Based Photocatalysts for Hydrogen Generation. *J. Phys. Chem. Lett.* **2013**, *4*, 753–759. [CrossRef]

57. Kim, H.-I.; Moon, G.-H.; Monllor-Satoca, D.; Park, Y.; Choi, W. Solar Photoconversion Using Graphene/TiO_2 Composites: Nanographene Shell on TiO_2 Core versus TiO_2 Nanoparticles on Graphene Sheet. *J. Phys. Chem. C* **2011**, *116*, 1535–1543. [CrossRef]

58. Wu, Z.; Zhou, Z.; Zhang, Y.; Wang, J.; Shi, H.; Shen, Q.; Wei, G.; Zhao, G. Simultaneous photoelectrocatalytic aromatic organic pollutants oxidation for hydrogen production promotion with a self-biasing photoelectrochemical cell. *Électrochim. Acta* **2017**, *254*, 140–147. [CrossRef]

59. Kim, B.J.; Lee, Y.S.; Park, S.J. Preparation of platinum-decorated porous graphite nanofibers, and their hydrogen storage behaviors. *J. Colloid Interface Sci.* **2008**, *318*, 530–533. [CrossRef]

60. Zhang, Y.; Park, S.-J. Fabrication and characterization of flower-like BiOI/Pt heterostructure with enhanced photocatalytic activity under visible light irradiation. *J. Solid State Chem.* **2017**, *253*, 421–429. [CrossRef]

61. Zhang, Y.; Park, S.-J. Incorporation of RuO_2 into charcoal-derived carbon with controllable microporosity by CO_2 activation for high-performance supercapacitor. *Carbon* **2017**, *122*, 287–297. [CrossRef]

62. Panthi, G.; Park, M.; Kim, H.-Y.; Park, S.-J. Electrospun polymeric nanofibers encapsulated with nanostructured materials and their applications: A review. *J. Ind. Eng. Chem.* **2015**, *24*, 1–13. [CrossRef]

63. Panthi, G.; Park, M.; Kim, H.-Y.; Lee, S.-Y.; Park, S.-J. Electrospun ZnO hybrid nanofibers for photodegradation of wastewater containing organic dyes: A review. *J. Ind. Eng. Chem.* **2015**, *21*, 26–35. [CrossRef]

64. Kim, S.; Park, S. Electroactivity of Pt–Ru/polyaniline composite catalyst-electrodes prepared by electrochemical deposition methods. *Solid State Ion.* **2008**, *178*, 1915–1921. [CrossRef]

65. Park, S.J.; Kim, B.J. Influence of oxygen plasma treatment on hydrogen chloride removal of activated carbon fibers. *J. Colloid Interface Sci.* **2004**, *275*, 590–595. [CrossRef] [PubMed]

66. Park, S.-J.; Kim, J.S. Modifications produced by electrochemical treatments oncarbon blacks Microstructures and mechanical interfacial properties. *Carbon* **2001**, *39*, 2011–2016. [CrossRef]

67. Chen, W.-T.; Chan, A.; Sun-Waterhouse, D.; Llorca, J.; Idriss, H.; Waterhouse, G.I.N. Performance comparison of Ni/TiO$_2$ and Au/TiO$_2$ photocatalysts for H$_2$ production in different alcohol-water mixtures. *J. Catal.* **2018**, *367*, 27–42. [CrossRef]

68. Hou, H.; Liu, H.; Gao, F.; Shang, M.; Wang, L.; Xu, L.; Wong, W.-Y.; Yang, W. Packaging BiVO$_4$ nanoparticles in ZnO microbelts for efficient photoelectrochemical hydrogen production. *Electrochim. Acta* **2018**, *283*, 497–508. [CrossRef]

69. Belhadj, H.; Hamid, S.; Robertson, P.K.J.; Bahnemann, D.W. Mechanisms of Simultaneous Hydrogen Production and Formaldehyde Oxidation in H$_2$O and D$_2$O over Platinized TiO$_2$. *ACS Catal.* **2017**, *7*, 4753–4758. [CrossRef]

70. Akbarzadeh, R.; Ghaedi, M.; Nasiri Kokhdan, S.; Jannesar, R.; Sadeghfar, F.; Sadri, F.; Tayebi, L. Electrochemical hydrogen storage, photocatalytical and antibacterial activity of Fe Ag bimetallic nanoparticles supported on TiO$_2$ nanowires. *Int. J. Hydrogen Energy* **2018**, *43*, 18316–18329. [CrossRef]

71. Chen, C.; Cai, W.M.; Long, M.C.; Zhou, B.X.; Wu, Y.H.; Wu, D.Y.; Feng, Y.J. Synthesis of Visible Light Responsive Graphene Oxide/ TiO$_2$ Composites with p/n Heterojunction. *ACS Nano* **2010**, *4*, 6425–6432. [CrossRef] [PubMed]

72. Cao, S.; Yu, J. g-C$_3$N$_4$-Based Photocatalysts for Hydrogen Generation. *J. Phys. Chem. Lett.* **2014**, *5*, 2101–2107. [CrossRef] [PubMed]

73. Rather, R.A.; Singh, S.; Pal, B. A C$_3$N$_4$ surface passivated highly photoactive Au-TiO$_2$ tubular nanostructure for the efficient H$_2$ production from water under sunlight irradiation. *Appl. Catal. B Environ.* **2017**, *213*, 9–17. [CrossRef]

74. Bian, H.; Ji, Y.; Yan, J.; Li, P.; Li, L.; Li, Y.; Frank Liu, S. In Situ Synthesis of Few-Layered g-C$_3$N$_4$ with Vertically Aligned MoS$_2$ Loading for Boosting Solar-to-Hydrogen Generation. *Small* **2018**, *14*, 1703003. [CrossRef] [PubMed]

75. Tian, N.; Zhang, Y.; Li, X.; Xiao, K.; Du, X.; Dong, F.; Waterhouse, G.I.N.; Zhang, T.; Huang, H. Precursor-reforming protocol to 3D mesoporous g-C$_3$N$_4$ established by ultrathin self-doped nanosheets for superior hydrogen evolution. *Nano Energy* **2017**, *38*, 72–81. [CrossRef]

76. Xu, X.; Si, Z.; Liu, L.; Wang, Z.; Chen, Z.; Ran, R.; He, Y.; Weng, D. CoMoS$_2$/rGO/C$_3$N$_4$ ternary heterojunctions catalysts with high photocatalytic activity and stability for hydrogen evolution under visible light irradiation. *Appl. Surf. Sci.* **2018**, *435*, 1296–1306. [CrossRef]

77. Luo, X.; Wu, Z.; Liu, Y.; Ding, S.; Zheng, Y.; Jiang, Q.; Zhou, T.; Hu, J. Engineering Amorphous Carbon onto Ultrathin g-C$_3$N$_4$ to Suppress Intersystem Crossing for Efficient Photocatalytic H$_2$ Evolution. *Adv. Mater. Interfaces* **2018**, *5*, 1800859. [CrossRef]

78. Wang, Y.; Wang, X.; Antonietti, M. Polymeric graphitic carbon nitride as a heterogeneous organocatalyst: From photochemistry to multipurpose catalysis to sustainable chemistry. *Angewandte Chem.* **2012**, *51*, 68–89. [CrossRef]

79. Wang, X.; Maeda, K.; Thomas, A.; Takanabe, K.; Xin, G.; Carlsson, J.M.; Domen, K.; Antonietti, M. A metal-free polymeric photocatalyst for hydrogen production from water under visible light. *Nat. Mater.* **2009**, *8*, 76–80. [CrossRef]

80. Gao, H.; Yang, H.; Xu, J.; Zhang, S.; Li, J. Strongly Coupled g-C$_3$N$_4$ Nanosheets-Co$_3$O$_4$ Quantum Dots as 2D/0D Heterostructure Composite for Peroxymonosulfate Activation. *Small* **2018**, *14*, 1801353. [CrossRef]

81. Chen, Z.; Xia, K.; She, X.; Mo, Z.; Zhao, S.; Yi, J.; Xu, Y.; Chen, H.; Xu, H.; Li, H. 1D metallic MoO$_2$-C as co-catalyst on 2D g-C$_3$N$_4$ semiconductor to promote photocatlaytic hydrogen production. *Appl. Surf. Sci.* **2018**, *447*, 732–739. [CrossRef]

82. Kong, L.; Ji, Y.; Dang, Z.; Yan, J.; Li, P.; Li, Y.; Liu, S.F. g-C$_3$N$_4$ Loading Black Phosphorus Quantum Dot for Efficient and Stable Photocatalytic H2 Generation under Visible Light. *Adv. Funct. Mater.* **2018**, *28*, 1800668. [CrossRef]

83. Wan, J.; Pu, C.; Wang, R.; Liu, E.; Du, X.; Bai, X.; Fan, J.; Hu, X. A facile dissolution strategy facilitated by H_2SO_4 to fabricate a 2D metal-free g-C_3N_4/rGO heterojunction for efficient photocatalytic H_2 production. *Int. J. Hydrogen Energy* **2018**, *43*, 7007–7019. [CrossRef]

84. Marcì, G.; García-López, E.I.; Palmisano, L. Polymeric carbon nitride (C_3N_4) as heterogeneous photocatalyst for selective oxidation of alcohols to aldehydes. *Catal. Today* **2018**, *315*, 126–137. [CrossRef]

85. Oh, W.-D.; Lok, L.-W.; Veksha, A.; Giannis, A.; Lim, T.-T. Enhanced photocatalytic degradation of bisphenol A with Ag-decorated S-doped g-C_3N_4 under solar irradiation: Performance and mechanistic studies. *Chem. Eng. J.* **2018**, *333*, 739–749. [CrossRef]

86. Hou, Y.; Laursen, A.B.; Zhang, J.; Zhang, G.; Zhu, Y.; Wang, X.; Dahl, S.; Chorkendorff, I. Layered nanojunctions for hydrogen-evolution catalysis. *Angewandte Chem.* **2013**, *52*, 3621–3625. [CrossRef] [PubMed]

87. Zhou, J.; Zhao, Y.; Bao, J.; Huo, D.; Fa, H.; Shen, X.; Hou, C. One-step electrodeposition of Au-Pt bimetallic nanoparticles on MoS_2 nanoflowers for hydrogen peroxide enzyme-free electrochemical sensor. *Electrochim. Acta* **2017**, *250*, 152–158. [CrossRef]

88. Yang, Y.; Gao, P.; Ren, X.; Sha, L.; Yang, P.; Zhang, J.; Chen, Y.; Yang, L. Massive Ti^{3+} self-doped by the injected electrons from external Pt and the efficient photocatalytic hydrogen production under visible-Light. *Appl. Catal. B Environ.* **2017**, *218*, 751–757. [CrossRef]

89. Fang, J.; Gu, J.; Liu, Q.; Zhang, W.; Su, H.; Zhang, D. Three-Dimensional CdS/Au Butterfly Wing Scales with Hierarchical Rib Structures for Plasmon-Enhanced Photocatalytic Hydrogen Production. *ACS Appl. Mater. Interfaces* **2018**, *10*, 19649–19655. [CrossRef] [PubMed]

90. Chang, Y.; Yu, K.; Zhang, C.; Yang, Z.; Feng, Y.; Hao, H.; Jiang, Y.; Lou, L.-L.; Zhou, W.; Liu, S. Ternary CdS/Au/3DOM-SrTiO$_3$ composites with synergistic enhancement for hydrogen production from visible-light photocatalytic water splitting. *Appl. Catal. B Environ.* **2017**, *215*, 74–84. [CrossRef]

91. Masudy-Panah, S.; Siavash Moakhar, R.; Chua, C.S.; Kushwaha, A.; Dalapati, G.K. Stable and Efficient CuO Based Photocathode through Oxygen-Rich Composition and Au-Pd Nanostructure Incorporation for Solar-Hydrogen Production. *ACS Appl. Mater. Interfaces* **2017**, *9*, 27596–27606. [CrossRef] [PubMed]

92. Ortiz, N.; Zoellner, B.; Hong, S.J.; Ji, Y.; Wang, T.; Liu, Y.; Maggard, P.A.; Wang, G. Harnessing Hot Electrons from Near IR Light for Hydrogen Production Using Pt-End-Capped-AuNRs. *ACS Appl. Mater. Interfaces* **2017**, *9*, 25962–25969. [CrossRef]

93. Wang, Y.; Zhao, J.; Li, Y.; Wang, C. Selective photocatalytic CO_2 reduction to CH_4 over Pt/In_2O_3: Significant role of hydrogen adatom. *Appl. Catal. B Environ.* **2018**, *226*, 544–553. [CrossRef]

94. Jiang, J.-Z.; Ren, L.-Q.; Huang, Y.-P.; Li, X.-D.; Wu, S.-H.; Sun, J.-J. 3D Nanoporous Gold-Supported Pt Nanoparticles as Highly Accelerating Catalytic Au-Pt Micromotors. *Adv. Mater. Interfaces* **2018**, *5*, 1701689. [CrossRef]

95. Lang, Q.; Chen, Y.H.; Huang, T.L.; Yang, L.N.; Zhong, S.X.; Wu, L.J.; Chen, J.R.; Bai, S. Graphene 'bridge' in transferring hot electrons from plasmonic Ag nanocubes to TiO_2 nanosheets for enhanced visible light photocatalytic hydrogen evolution. *Appl. Catal. B Environ.* **2018**, *220*, 182–190. [CrossRef]

96. Zhang, Y.; Park, S.-J. Facile construction of MoO_3@ZIF-8 core-shell nanorods for efficient photoreduction of aqueous Cr (VI). *Appl. Catal. B Environ.* **2019**, *240*, 92–101. [CrossRef]

97. Kamijyo, K.; Takashima, T.; Yoda, M.; Osaki, J.; Irie, H. Facile synthesis of a red light-inducible overall water-splitting photocatalyst using gold as a solid-state electron mediator. *Chem. Commun.* **2018**, *54*, 7999–8002. [CrossRef] [PubMed]

98. Maeda, K.; Domen, K. Photocatalytic Water Splitting: Recent Progress and Future Challenges. *J. Phys. Chem. Lett.* **2010**, *1*, 2655–2661. [CrossRef]

99. Iwase, A.; Ng, Y.H.; Ishiguro, Y.; Kudo, A.; Amal, R. Reduced graphene oxide as a solid-state electron mediator in Z-scheme photocatalytic water splitting under visible light. *J. Am. Chem. Soc.* **2011**, *133*, 11054–11057. [CrossRef] [PubMed]

100. Zhang, Y.; Park, S.-J. Formation of hollow MoO_3/SnS_2 heterostructured nanotubes for efficient light-driven hydrogen peroxide production. *J. Mater. Chem. A* **2018**, *6*, 20304–20312. [CrossRef]

101. Xu, C.; Qiu, P.; Li, L.; Chen, H.; Jiang, F.; Wang, X. Bismuth Subcarbonate with Designer Defects for Broad-Spectrum Photocatalytic Nitrogen Fixation. *ACS Appl. Mater. Interfaces* **2018**, *10*, 25321–25328. [CrossRef] [PubMed]

102. Kobayashi, R.; Takashima, T.; Tanigawa, S.; Takeuchi, S.; Ohtani, B.; Irie, H. A heterojunction photocatalyst composed of zinc rhodium oxide, single crystal-derived bismuth vanadium oxide, and silver for overall pure-water splitting under visible light up to 740 nm. *Phys. Chem. Chem. Phys. PCCP* **2016**, *18*, 27754–27760. [CrossRef] [PubMed]

103. Iwashina, K.; Iwase, A.; Ng, Y.H.; Amal, R.; Kudo, A. Z-schematic water splitting into H_2 and O_2 using metal sulfide as a hydrogen-evolving photocatalyst and reduced graphene oxide as a solid-state electron mediator. *J. Am. Chem. Soc.* **2015**, *137*, 604–607. [CrossRef] [PubMed]

104. Yu, X.; Liu, G.; Li, W.; An, L.; Li, Z.; Liu, J.; Hu, P. Mesocrystalline Ta_2O_5 nanosheets supported Pd Pt nanoparticles for efficient photocatalytic hydrogen production. *Int. J. Hydrogen Energy* **2018**, *43*, 8232–8242. [CrossRef]

105. Low, J.; Yu, J.; Jaroniec, M.; Wageh, S.; Al-Ghamdi, A.A. Heterojunction Photocatalysts. *Adv. Mater.* **2017**, *29*, 1601694. [CrossRef] [PubMed]

106. Zhu, M.; Zhai, C.; Fujitsuka, M.; Majima, T. Noble metal-free near-infrared-driven photocatalyst for hydrogen production based on 2D hybrid of black Phosphorus/WS_2. *Appl. Catal. B Environ.* **2018**, *221*, 645–651. [CrossRef]

107. Do, J.Y.; Lee, J.H.; Park, N.-K.; Lee, T.J.; Lee, S.T.; Kang, M. Synthesis and characterization of $Ni_2-xPd_xMnO_4/\gamma-Al_2O_3$ catalysts for hydrogen production via propane steam reforming. *Chem. Eng. J.* **2018**, *334*, 1668–1678. [CrossRef]

108. Qin, Z.; Xue, F.; Chen, Y.; Shen, S.; Guo, L. Spatial charge separation of one-dimensional Ni_2P $Cd_{0.9}Zn_{0.1}S/g-C_3N_4$ heterostructure for high-quantum-yield photocatalytic hydrogen production. *Appl. Catal. B Environ.* **2017**, *217*, 551–559. [CrossRef]

109. Hu, Z.; Wang, X.; Dong, H.; Li, S.; Li, X.; Li, L. Efficient photocatalytic degradation of tetrabromodiphenyl ethers and simultaneous hydrogen production by TiO_2-Cu_2O composite films in N_2 atmosphere: Influencing factors, kinetics and mechanism. *J. Hazard. Mater.* **2017**, *340*, 1–15. [CrossRef]

110. Iervolino, G.; Vaiano, V.; Sannino, D.; Rizzo, L.; Galluzzi, A.; Polichetti, M.; Pepe, G.; Campiglia, P. Hydrogen production from glucose degradation in water and wastewater treated by Ru-$LaFeO_3$/Fe_2O_3 magnetic particles photocatalysis and heterogeneous photo-Fenton. *Int. J. Hydrogen Energy* **2018**, *43*, 2184–2196. [CrossRef]

111. Imran, M.; Yousaf, A.B.; Kasak, P.; Zeb, A.; Zaidi, S.J. Highly efficient sustainable photocatalytic Z-scheme hydrogen production from an α-Fe_2O_3 engineered ZnCdS heterostructure. *J. Catal.* **2017**, *353*, 81–88. [CrossRef]

112. Subha, N.; Mahalakshmi, M.; Myilsamy, M.; Neppolian, B.; Murugesan, V. Direct Z-scheme heterojunction nanocomposite for the enhanced solar H_2 production. *Appl. Catal. A Gen.* **2018**, *553*, 43–51. [CrossRef]

113. Zhang, J.; Yan, W.; An, Z.; Song, H.; He, J. Interface–Promoted Dehydrogenation and Water–Gas Shift toward High-Efficient H_2 Production from Aqueous Phase Reforming of Cellulose. *ACS Sustain. Chem. Eng.* **2018**, *6*, 7313–7324. [CrossRef]

114. Vinodgopal, K.; Kamat, P.V. Enhanced rates of photocatalytic degradation of an azo dye using SnO_2/TiO_2 coupled semiconductor thin films. *Environ. Sci. Technol.* **1995**, *29*, 841–845. [CrossRef] [PubMed]

115. Ranjit, K.; Viswanathan, B. Synthesis, characterization and photocatalytic properties of iron-doped TiO_2 catalysts. *J. Photochem. Photobiol. A* **1997**, *108*, 79–84. [CrossRef]

116. Monai, M.; Montini, T.; Fonda, E.; Crosera, M.; Delgado, J.J.; Adami, G.; Fornasiero, P. Nanostructured Pd Pt nanoparticles: Evidences of structure/performance relations in catalytic H_2 production reactions. *Appl. Catal. B Environ.* **2018**, *236*, 88–98. [CrossRef]

117. Wang, Q.; He, J.; Shi, Y.; Zhang, S.; Niu, T.; She, H.; Bi, Y.; Lei, Z. Synthesis of MFe_2O_4 (M = Ni, Co)/$BiVO_4$ film for photolectrochemical hydrogen production activity. *Appl. Catal. B Environ.* **2017**, *214*, 158–167. [CrossRef]

118. Xu, J.; Gao, J.; Qi, Y.; Wang, C.; Wang, L. Anchoring Ni_2P on the UiO-66-NH_2/$g-C_3N_4$-derived C-doped ZrO_2/$g-C_3N_4$ Heterostructure: Highly Efficient Photocatalysts for H_2 Production from Water Splitting. *ChemCatChem* **2018**, *10*, 3327–3335. [CrossRef]

119. Wang, Z.; Jin, Z.; Wang, G.; Ma, B. Efficient hydrogen production over MOFs (ZIF-67) and $g-C_3N_4$ boosted with MoS_2 nanoparticles. *Int. J. Hydrogen Energy* **2018**, *43*, 13039–13050. [CrossRef]

120. Fu, J.; Zhu, B.; You, W.; Jaroniec, M.; Yu, J. A flexible bio-inspired H_2- production photocatalyst. *Appl. Catal. B Environ.* **2018**, *220*, 148–160. [CrossRef]

121. Zhang, S.; Liu, X.; Liu, C.; Luo, S.; Wang, L.; Cai, T.; Zeng, Y.; Yuan, J.; Dong, W.; Pei, Y.; et al. MoS$_2$ Quantum Dot Growth Induced by S Vacancies in a ZnIn$_2$S$_4$ Monolayer: Atomic-Level Heterostructure for Photocatalytic Hydrogen Production. *ACS Nano* **2018**, *12*, 751–758. [CrossRef] [PubMed]

122. Li, K.; Gao, S.; Wang, Q.; Xu, H.; Wang, Z.; Huang, B.; Dai, Y.; Lu, J. In-Situ-Reduced Synthesis of Ti(3)(+) Self-Doped TiO$_{(2)}$/g-C$_{(3)}$N$_{(4)}$ Heterojunctions with High Photocatalytic Performance under LED Light Irradiation. *ACS Appl. Mater. Interfaces* **2015**, *7*, 9023–9030. [CrossRef] [PubMed]

123. Hong, S.J.; Lee, S.; Jang, J.S.; Lee, J.S. Heterojunction BiVO$_4$/WO$_3$ electrodes for enhanced photoactivity of water oxidation. *Energy Environ. Sci.* **2011**, *4*, 1781–1787. [CrossRef]

124. Huang, L.; Xu, H.; Li, Y.; Li, H.; Cheng, X.; Xia, J.; Xu, Y.; Cai, G. Visible-light-induced WO$_3$/g-C$_3$N$_4$ composites with enhanced photocatalytic activity. *Dalton Trans.* **2013**, *42*, 8606–8616. [CrossRef] [PubMed]

125. Pan, C.; Xu, J.; Wang, Y.; Li, D.; Zhu, Y. Dramatic Activity of C$_3$N$_4$/BiPO$_4$ Photocatalyst with Core/Shell Structure Formed by Self-Assembly. *Adv. Funct. Mater.* **2012**, *22*, 1518–1524. [CrossRef]

126. Zhang, J.; Qiao, S.Z.; Qi, L.; Yu, J. Fabrication of NiS modified CdS nanorod p-n junction photocatalysts with enhanced visible-light photocatalytic H$_2$-production activity. *Phys. Chem. Chem. Phys. PCCP* **2013**, *15*, 12088–12094. [CrossRef]

127. Heo, Y.-J.; Zhang, Y.; Rhee, K.Y.; Park, S.-J. Synthesis of PAN/PVDF nanofiber composites-based carbon adsorbents for CO$_2$ capture. *Compos. Part B Eng.* **2019**, *156*, 95–99. [CrossRef]

128. Zhang, Y.; Park, M.; Kim, H.Y.; Park, S.J. Moderated surface defects of Ni particles encapsulated with NiO nanofibers as supercapacitor with high capacitance and energy density. *J. Colloid Interface Sci.* **2017**, *500*, 155–163. [CrossRef]

MDPI

St. Alban-Anlage 66

4052 Basel

Switzerland

Tel. +41 61 683 77 34

Fax +41 61 302 89 18

www.mdpi.com

Catalysts Editorial Office

E-mail: catalysts@mdpi.com

www.mdpi.com/journal/catalysts